高等职业院校互联网+新形态创新系列教材·计算机系列

# Java Web 应用开发——
# SpringBoot+MyBatis+Spring+SpringMVC
# (微课版)

齐 洋 王 黎 原变青 主 编
杨 婷 段炬霞 王鹏成 郭俊杰 副主编

清华大学出版社
北京

## 内 容 简 介

本书是一本专为 Java Web 开发者设计的完整教程，涵盖了 SpringBoot、MyBatis、Spring 和 SpringMVC 技术栈的诸多方面。本书从基础知识到高级技巧、从理论概念到实际应用，为读者提供了一条完整的学习路径。

本书共 10 章。第 1 章主要介绍 SpringBoot 开发环境的配置。第 2~5 章主要讲解 MyBatis、Spring 和 SpringMVC 框架的开发，其中 MyBatis 部分包含 MyBatis 的配置、基本数据操作、动态 SQL、关联映射等，Spring 部分包含 Spring 关键技术，如控制反转(IoC)和面向切面的编程(AOP)等，SpringMVC 部分包含知识点简介与基于注解的开发流程等。第 6 章主要讲解上述三大框架的整合过程。第 7~8 章主要讲解 SpringBoot 及其与 MyBatis 框架的整合开发流程。第 9 章主要讲解 Java Web 开发中的一些常用功能，如过滤器、拦截器、文件上传和缓存等。第 10 章通过一个教学信息管理系统案例，讲解实际开发中 MyBatis、Spring、SpringMVC 和 SpringBoot 的应用。

为方便学习，读者通过扫描书中的二维码即可观看微课视频、动画讲解，扫描前言末尾左侧的二维码可下载源代码、习题答案等配套资源；针对教师，本书提供教学课件、教学大纲、试卷等资源，教师可扫描前言末尾右侧的二维码获取相关教学资源服务。

本书既可以作为高职高专院校计算机等相关专业的教学用书，也可以作为 Java Web 开发专业人员的培训参考用书。

本书封面贴有清华大学出版社防伪标签，无标签者不得销售。
版权所有，侵权必究。举报：010-62782989，beiqinquan@tup.tsinghua.edu.cn。

**图书在版编目(CIP)数据**

Java Web 应用开发：SpringBoot+MyBatis+Spring+SpringMVC：微课版 / 齐洋，王黎，原变青主编．
北京：清华大学出版社，2025.2. -- (高等职业院校互联网+新形态创新系列教材). -- ISBN 978-7-302-68321-6
Ⅰ．TP312.8
中国国家版本馆 CIP 数据核字第 20252UJ218 号

责任编辑：桑任松
封面设计：杨玉兰
责任校对：翟维维
责任印制：曹婉颖

出版发行：清华大学出版社
网　　址：https://www.tup.com.cn, https://www.wqxuetang.com
地　　址：北京清华大学学研大厦 A 座　邮　编：100084
社 总 机：010-83470000　邮　购：010-62786544
投稿与读者服务：010-62776969, c-service@tup.tsinghua.edu.cn
质量反馈：010-62772015, zhiliang@tup.tsinghua.edu.cn
课件下载：https://www.tup.com.cn, 010-62791865

印 装 者：北京鑫海金澳胶印有限公司
经　　销：全国新华书店
开　　本：185mm×260mm　印　张：17　字　数：414 千字
版　　次：2025 年 3 月第 1 版　印　次：2025 年 3 月第 1 次印刷
定　　价：52.00 元

产品编号：104830-01

# 前　　言

　　党的二十大报告中强调，我国要构建包含新一代信息技术、人工智能在内的新增长引擎。报告同时强调要完善科技创新体系，坚持创新在我国现代化建设全局中的核心地位，健全新型举国体制，强化国家战略科技力量，提升国家创新体系整体效能，形成具有全球竞争力的开放创新生态。这为我国新一代信息技术产业发展与信息技术的教育事业指明了方向。

　　随着新一代信息技术特别是互联网技术的迅猛发展和企业应用的持续升级，Java Web 应用开发已经成为软件行业中的一个重要领域。尤其是在微服务、分布式系统、云计算等技术趋势推动下，一个高效、可扩展、易维护的后台系统是每一个现代企业所亟需的。

　　SpringBoot、MyBatis、Spring 和 SpringMVC 是目前 Java Web 开发中非常流行和实用的技术组合。它们各自在框架设计、开发效率和数据访问等领域都有独到之处，组合在一起，可以为开发者提供一套强大而完整的开发工具集。

　　编写本书的初衷，是为了帮助那些对 Java Web 开发感兴趣的初学者和开发者，从零开始，系统掌握这几个框架的应用。无论是初学者，还是有一定经验的开发者，本书都会为你提供结构清晰、案例丰富、实战指导详尽的内容。

　　在本教材中，我们从基础知识出发，逐步深入每一个技术的核心部分，同时对于每一个知识点，都结合真实的项目案例，从功能设计到代码实现，每一步都有具体详细的介绍，帮助读者将所涉及的理论知识与其实际应用紧密结合，逐步熟练地掌握书中各个知识点的开发技术。我们也会探讨一些常见的开发问题、最佳实践和解决方案，帮助读者更好地理解和掌握这些技术。

　　本书共 10 章，具体内容如下。

　　第 1 章介绍了 SpringBoot 开发环境的详细搭建过程，相关软件工具的安装和配置，包含 Java 开发环境 JDK、依赖管理工具 Maven、开发工具 IDEA、数据库软件 MySQL 及连接工具 Navicat 的详细安装过程。

　　第 2 章介绍了 Java 的持久层框架 MyBatis 的概念、初步配置与简单开发过程，包含数据库的初始化、MyBatis 配置文件的编写、MyBatis 数据操作(增、删、改、查)代码的编写与简单测试。

　　第 3 章介绍了 Mybatis 高级应用，包含动态 SQL、数据关系一对一、一对多、多对多映射等高级功能。

　　第 4 章介绍了 Spring 框架的基本概念和常见的用法，包含控制反转(IoC)与面向切面编程(AOP)。

　　第 5 章介绍了 SpringMVC 框架的基础开发，对 SpringMVC 常用的注解都给出了相应的实例，并介绍了 JSON 和接口测试工具 Postman 的使用。

　　第 6 章结合实际项目介绍了 Spring、SpringMVC 和 MyBatis 三个框架的整合过程，给出了详细的开发和测试流程。

　　第 7 章在前述所有章节的基础上介绍了 SpringBoot 项目的详细配置和开发内容。

第 8 章介绍了 SpringBoot 和 MyBatis 的整合流程及实用工具 MyBatisPlus 的使用方法。

第 9 章介绍了 Java Web 开发中一些实用的功能与工具的使用，如请求过滤器、拦截器、文件上传和缓存的使用。

第 10 章为项目实战，综合了本书前面所有章节的内容，以一个教学信息管理系统为例，带领读者完成一个功能完备的实际项目的设计与开发。

本书由北京经济管理职业学院齐洋、王黎、原变青任主编，由北京经济管理职业学院杨婷、段炬霞以及东誉（北京）国际电子商务技术有限公司王鹏成、郭俊杰任副主编。由于编者水平有限，尽管对本书内容设计与结构安排进行了反复的斟酌和修正，但仍然难免存在不当之处，敬请各位专家和广大读者批评、指正。

最后，希望本书能够成为读者 Java Web 学习和实践过程中的得力助手，期待读者阅读后能够对 Java Web 应用开发有一个更深入、更全面的了解。

编　者

读者资源下载

教师资源服务

# 目　　录

## 第 1 章　SpringBoot 开发环境准备 ................................................................1

### 1.1　JDK 的安装 ..........................................................................................2
#### 1.1.1　下载 JDK ...................................................................................3
#### 1.1.2　安装 JDK ...................................................................................3
#### 1.1.3　配置环境变量 ............................................................................4
#### 1.1.4　验证安装 ...................................................................................5

### 1.2　Maven 的安装与配置 ...........................................................................6
#### 1.2.1　Maven 简介 ...............................................................................6
#### 1.2.2　下载 Maven ...............................................................................6
#### 1.2.3　解压文件 ...................................................................................7
#### 1.2.4　配置环境变量 ............................................................................7
#### 1.2.5　验证安装 ...................................................................................8
#### 1.2.6　修改 Maven 配置文件 ...............................................................8

### 1.3　IntelliJ IDEA 的安装 ............................................................................9
#### 1.3.1　下载 IntelliJ IDEA 安装包 .......................................................10
#### 1.3.2　安装 IntelliJ IDEA ...................................................................10
#### 1.3.3　在 IntelliJ IDEA 中设置 Maven ..............................................12

### 1.4　MySQL 数据库的安装 .......................................................................13
#### 1.4.1　MySQL 的安装 ........................................................................13
#### 1.4.2　Navicat 客户端的安装与使用 .................................................19
#### 1.4.3　MySQL 和 Navicat 的简单使用 ..............................................21

### 1.5　第一个 SpringBoot 程序 ....................................................................23
#### 1.5.1　添加依赖 .................................................................................23
#### 1.5.2　创建启动类和控制器 ..............................................................25
#### 1.5.3　测试 .........................................................................................27

### 本章小结 .......................................................................................................27
### 课后习题 .......................................................................................................28

## 第 2 章　MyBatis 框架初体验 .........................................................................29

### 2.1　MyBatis 介绍 ......................................................................................30
#### 2.1.1　MyBatis 概述 ...........................................................................30
#### 2.1.2　为什么使用 MyBatis ...............................................................30
#### 2.1.3　MyBatis 和其他 ORM 框架的对比 .........................................31

### 2.2　搭建 MyBatis 开发环境 .....................................................................31
#### 2.2.1　初始化数据库 ..........................................................................31

2.2.2 创建项目，添加依赖 .................................................. 32
2.2.3 mybatis-config.xml 文件 ........................................... 33
2.2.4 创建实体类 ........................................................ 34
2.2.5 创建 Mapper 接口 ................................................. 34
2.2.6 创建 Mapper 文件 ................................................. 35
2.2.7 创建测试类查询全部客户 ............................................ 35
2.3 MyBatis 增删改查 .......................................................... 37
2.3.1 查询单个客户 ...................................................... 37
2.3.2 插入客户 .......................................................... 38
2.3.3 删除客户 .......................................................... 39
2.3.4 修改客户 .......................................................... 40
本章小结 ..................................................................... 42
课后习题 ..................................................................... 42

## 第 3 章 深入使用 MyBatis 框架 .................................................. 45

3.1 动态 SQL ................................................................. 46
3.1.1 <if>、<where>标签 ................................................. 47
3.1.2 <choose>、<when>和<otherwise>标签 .............................. 48
3.1.3 <set>标签 ......................................................... 50
3.1.4 <foreach>标签 ..................................................... 51
3.2 MyBatis 关联映射 .......................................................... 53
3.2.1 一对一 ............................................................ 53
3.2.2 一对多 ............................................................ 56
3.2.3 多对多 ............................................................ 58
本章小结 ..................................................................... 61
课后习题 ..................................................................... 61

## 第 4 章 Spring 框架使用指南 .................................................... 63

4.1 Spring 介绍 ............................................................... 64
4.1.1 Spring 概念 ........................................................ 64
4.1.2 Spring 的特点 ...................................................... 64
4.2 Spring 的 IoC 和 DI ........................................................ 65
4.2.1 IoC、DI 案例 ...................................................... 65
4.2.2 Bean 的作用域 ..................................................... 68
4.2.3 Spring 基于注解开发 ............................................... 70
4.3 Spring AOP 案例 ........................................................... 71
本章小结 ..................................................................... 74
课后习题 ..................................................................... 74

## 第 5 章 SpringMVC 上手开发 .................................................... 77

5.1 SpringMVC 介绍 ........................................................... 78

5.2 搭建 SpringMVC 开发环境 .................................................. 79
5.3 Postman 工具 .................................................................. 85
5.4 JSON 简介 ...................................................................... 88
5.5 请求与响应注解 ............................................................... 89
    5.5.1 @RequestMapping 注解 ............................................. 89
    5.5.2 @RequestParam 注解 ................................................. 91
    5.5.3 @ResponseBody 注解 ................................................. 93
    5.5.4 @GetMapping 注解 .................................................... 94
    5.5.5 @RestController 注解 ................................................ 95
    5.5.6 @RequestBody 注解 ................................................... 95
本章小结 ............................................................................... 96
课后习题 ............................................................................... 96

## 第 6 章  SSM 整合开发 ............................................................ 99

6.1 搭建 SSM 基础环境 ......................................................... 100
    6.1.1 创建 Maven 项目 ..................................................... 100
    6.1.2 Spring 整合 MyBatis ................................................. 102
    6.1.3 Spring 整合 SpringMVC ............................................ 105
6.2 功能模块开发 ................................................................. 107
    6.2.1 数据层开发 ............................................................ 107
    6.2.2 业务层开发 ............................................................ 109
    6.2.3 控制器层开发 ......................................................... 110
6.3 接口测试 ....................................................................... 112
本章小结 ............................................................................. 116
课后习题 ............................................................................. 116

## 第 7 章  详解 SpringBoot ...................................................... 119

7.1 SpringBoot 的配置 .......................................................... 120
    7.1.1 SpringBoot 依赖说明 ................................................ 120
    7.1.2 SpringBoot 核心注解 ................................................ 121
7.2 YAML 配置文件 .............................................................. 124
    7.2.1 语法规则 ............................................................... 124
    7.2.2 SpringBoot 属性配置 ................................................ 125
    7.2.3 SpringBoot 多环境配置 ............................................. 126
7.3 SpringBoot 单元测试 ....................................................... 127
    7.3.1 创建 BookService 接口和实现类 .................................. 127
    7.3.2 创建测试类 ........................................................... 128
本章小结 ............................................................................. 129
课后习题 ............................................................................. 129

## 第 8 章　SpringBoot 集成 MyBatis .................................................. 131

### 8.1　环境准备 ............................................................................. 132
### 8.2　功能开发 ............................................................................. 135
### 8.3　接口测试 ............................................................................. 138
### 8.4　MyBatisPlus 简介与应用 ................................................... 139
#### 8.4.1　MyBatisPlus 简介 ........................................................ 139
#### 8.4.2　MyBatisPlus 的简单使用 ............................................ 139
### 本章小结 ..................................................................................... 144
### 课后习题 ..................................................................................... 144

## 第 9 章　过滤器、拦截器、文件上传和缓存 ................................. 147

### 9.1　过滤器 ................................................................................. 148
### 9.2　拦截器 ................................................................................. 152
### 9.3　文件上传 ............................................................................. 155
### 9.4　SpringBoot 整合 Redis ....................................................... 156
#### 9.4.1　Redis 简介 .................................................................... 156
#### 9.4.2　添加 Redis Maven 依赖 .............................................. 157
#### 9.4.3　下载与启动 Redis 服务端、客户端 .......................... 158
#### 9.4.4　编写 Redis 使用代码 .................................................. 159
#### 9.4.5　单元测试 ...................................................................... 160
### 本章小结 ..................................................................................... 161
### 课后习题 ..................................................................................... 161

## 第 10 章　项目实战：教学信息管理系统 ..................................... 163

### 10.1　系统概述 ........................................................................... 164
#### 10.1.1　系统功能介绍 ............................................................ 164
#### 10.1.2　系统后端架构设计 .................................................... 165
#### 10.1.3　文件组织结构 ............................................................ 165
### 10.2　数据库设计 ....................................................................... 166
### 10.3　搭建环境 ........................................................................... 168
#### 10.3.1　前端系统环境搭建 .................................................... 168
#### 10.3.2　后端系统环境搭建 .................................................... 172
### 10.4　系统基础功能 ................................................................... 177
#### 10.4.1　解决项目跨域问题 .................................................... 177
#### 10.4.2　封装 Controller 的响应 ............................................. 178
#### 10.4.3　封装分页查询参数与结果 ........................................ 180
### 10.5　管理员模块 ....................................................................... 182
#### 10.5.1　管理员登录、登出和获取信息 ................................ 182
#### 10.5.2　管理员访问验证 ........................................................ 196

10.6　教学信息管理模块 ......199
　　10.6.1　学院管理 ......199
　　10.6.2　班级管理 ......212
　　10.6.3　学生管理 ......223
　　10.6.4　课程管理 ......236
　　10.6.5　成绩管理 ......247
本章小结 ......261
课后习题 ......261

# 参考文献 ......262

# 第 1 章
# SpringBoot 开发环境准备

**学习目标**

1. JDK、Maven、IntelliJ IDEA、MySQL 数据库的安装与配置。
2. 掌握如何搭建 SpringBoot 的开发环境。

**学习要点**

1. 了解 JDK 的功能和重要性,掌握 JDK 的安装步骤和环境变量配置。
2. 掌握 IntelliJ IDEA 安装与初步使用,能进行基本的项目管理操作,如项目创建。
3. 能够安装 Maven,并在 IntelliJ IDEA 中进行配置。
4. 掌握 MySQL 数据库的安装步骤。
5. 了解基本的数据库操作,如创建数据库、创建表、插入和查询数据等。
6. 使用 IntelliJ IDEA 快速创建 SpringBoot 项目。

本章知识点结构如图 1-1 所示。

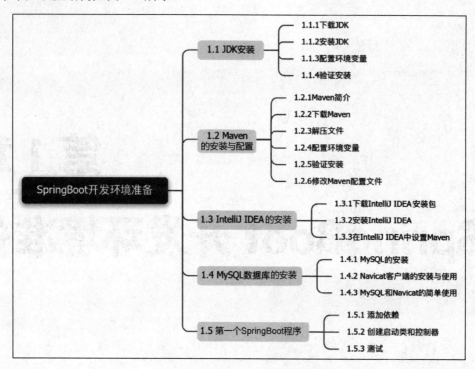

图 1-1　SpringBoot 开发环境准备

SpringBoot 是一个开源的 Java 框架，专为简化 Spring 应用程序的创建、配置和部署而设计。它是 Spring 框架的一个扩展，旨在帮助开发者更快速、更轻松地开发基于 Spring 的应用程序。以下是 SpringBoot 的几个关键特点。

(1) 快速开发：SpringBoot 使得创建 Spring 应用变得快捷和容易。它提供了大量预设配置模块，减少了烦琐的配置工作。

(2) 约定优于配置：SpringBoot 采用"约定优于配置"的原则，提供了合理的默认配置，同时也允许开发者在必要时进行定制。

(3) 独立运行：SpringBoot 应用可以打包成单一的、可独立运行的 Jar 文件，其中包括了所有必要的依赖、类和资源。这样的打包方式简化了部署过程。

(4) 内嵌服务器：SpringBoot 应用内嵌了 Tomcat、Jetty 或 Undertow 服务器，无须单独部署应用服务器。

(5) 自动配置：SpringBoot 自动配置项目的各个方面，如数据源、MVC 框架等，它基于项目中添加的依赖。

## 1.1　JDK 的安装

JDK (Java Development Kit) 是 Java 语言的官方开发工具包。它为 Java 开发人员提供了一系列的工具、可执行文件、调试器和库，以支持 Java 程序的开发和运行。本书使用当前应用最广泛的 JDK8 版本，下面讲解其具体的下载、安装和配置步骤。

## 1.1.1 下载 JDK

打开 Oracle 官方网站下载页面，网址如下：
https://www.oracle.com/java/technologies/javase/javase8-archive-downloads.html
选择对应的 Windows 平台版本进行下载。

微课：JDK 软件下载

## 1.1.2 安装 JDK

安装 JDK 的具体步骤如下。

（1）双击下载的安装文件，打开 JDK 的欢迎安装界面，单击"下一步"按钮，如图 1-2 所示。

微课：JDK 的安装

（2）选择安装路径，默认安装在 C:\Program Files\Java\jdk1.8.0_201 目录，单击"下一步"按钮，如图 1-3 所示。

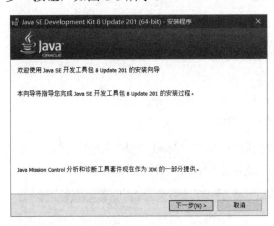

图 1-2 JDK 的欢迎安装界面

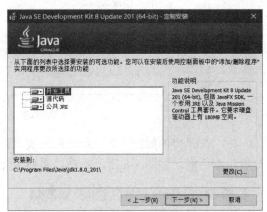

图 1-3 选择安装路径

（3）选择 JRE 的安装路径，默认安装在 C:\Program Files\Java\jre1.8.0_201 目录，单击"下一步"按钮，如图 1-4 所示。

（4）提示安装成功，单击"关闭"按钮，如图 1-5 所示。

图 1-4 选择 JRE 安装路径

图 1-5 安装成功提示界面

### 1.1.3 配置环境变量

JDK 安装成功之后，还需要在操作系统中配置对应的环境变量才能够正常使用。配置环境变量的具体操作步骤如下。

（1）在计算机桌面右击"计算机"或"此电脑"图标，在弹出的快捷菜单中选择"属性"命令，在打开的"系统"窗口中单击"高级系统设置"链接，如图1-6所示。

图1-6 "系统"窗口

（2）在弹出的如图1-7所示"系统属性"对话框中单击"环境变量"按钮，弹出"环境变量"对话框，如图1-8所示。在"系统变量"选项组中单击"新建"按钮。

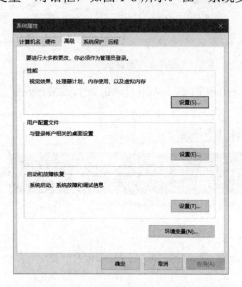

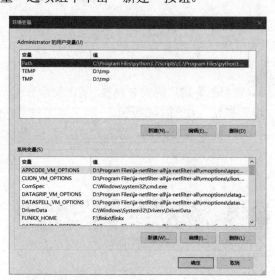

图1-7 "系统属性"对话框　　　　图1-8 "环境变量"对话框

（3）弹出 "编辑系统变量"对话框，在"变量名"和"变量值"文本框中分别输入"JAVA_HOME"和"C:\Program Files\Java\jdk1.8.0_201"，然后单击"确定"按钮，如图1-9所示。

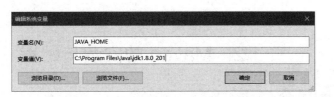

图 1-9 "编辑系统变量"对话框

(4) JAVA_HOME 配置好之后,将%JAVA_HOME%\bin 加入"系统变量"列表框的 Path 中。选择已经存在的 Path 变量,单击"编辑"按钮,如图 1-10 所示。

(5) 弹出如图 1-11 所示的"编辑环境变量"对话框,单击"新建"按钮,在下方输入框中输入"%JAVA_HOME%\bin",然后单击"确定"按钮,如图 1-11 所示。

图 1-10 编辑 Path 环境变量

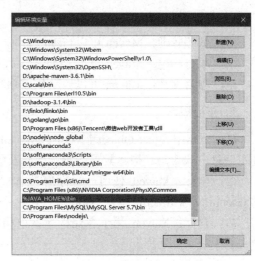

图 1-11 新建 Path 环境变量

## 1.1.4 验证安装

安装 JDK 与配置环境变量之后,需要验证安装与配置的正确性,以确保 JDK 可以正确使用。验证安装的具体操作步骤如下。

(1) 按 Win+R 快捷键,打开"运行"对话框,输入"cmd"以打开 CMD 命令行界面,如图 1 12 所示。

(2) 在 CMD 命令行界面中输入"java -version"。如果看到类似 java version "1.8.0_201"的输出,就表示安装成功了,如图 1-13 所示。

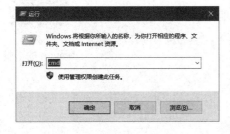

图 1-12 "运行"对话框

图 1-13 安装成功

## 1.2 Maven 的安装与配置

### 1.2.1 Maven 简介

Maven 是一款开源的项目管理和构建自动化工具,主要服务于 Java 平台,但也可用于构建和管理基于 C#、Ruby 等其他语言的项目。Maven 使用了一个统一的构建系统,所以开发者可以轻松地管理项目,而无论项目大小。

动画:Maven 在软件开发中的重要性与优势

**1. 主要功能**

(1) 依赖管理:Maven 可以自动处理项目依赖,减少手动导入依赖的麻烦。
(2) 项目构建:Maven 可以对 Java 源代码进行编译、测试和打包。
(3) 项目文档:Maven 可以自动生成项目文档。

**2. 核心概念**

(1) POM(Project Object Model):Maven 的基本工作单位,用 XML 文件描述了项目的基本信息、源代码、构建配置、测试代码及配置等方面的信息。
(2) 坐标:Maven 使用一组称为"坐标"的标识符来定位项目和依赖项,这些坐标包括 GroupId、ArtifactId 和 Version。
(3) 仓库:Maven 用于存储构建输出和第三方依赖的地方,可以是本地的,也可以是中央仓库或其他远程仓库。

**3. Maven 生命周期**

Maven 的生命周期定义了构建过程的几个阶段,如 clean、validate、compile、test、package、install、deploy 等,这些阶段以一定的顺序执行。

**4. 使用 Maven**

要使用 Maven,我们需要在项目中创建一个 pom.xml 文件,该文件包含了项目的基本信息、依赖、插件等。这样,Maven 就可以根据 pom.xml 中的信息来管理和构建项目。

### 1.2.2 下载 Maven

本书中使用 Maven 3.6.1 版本,使用浏览器访问网址 http://archive.apache.org/dist/maven/ maven-3/3.6.1/binaries,单击 apache-maven-3.6.1-bin.zip 链接进行下载,如图 1-14 所示。

微课:Maven 的下载

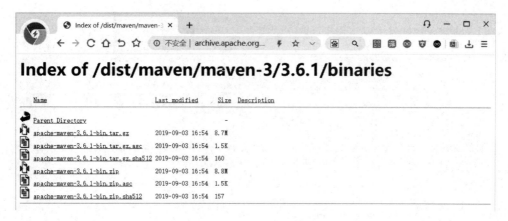

图 1-14　Maven 下载页面

## 1.2.3　解压文件

微课：Maven
软件安装与配置

Maven 不需要安装，直接解压缩和配置环境变量就可以使用。使用任意解压软件(如 WinRAR、7-Zip 等)解压文件到指定目录，如 D:\apache-maven-3.6.1，如图 1-15 所示。

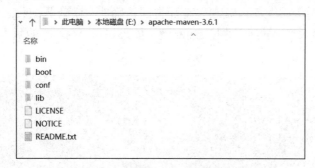

图 1-15　将 Maven 解压到指定目录

## 1.2.4　配置环境变量

配置 Maven 环境变量的具体操作步骤如下。

(1) 在计算机桌面右击"计算机"或"此电脑"图标，在弹出的快捷菜单中选择"属性"命令。在打开的"系统"窗口左侧单击"高级系统设置"链接。在弹出的"系统属性"对话框中，单击"环境变量"按钮，弹出"环境变量"对话框。在"系统变量"选项组中，单击"新建"按钮。

创建一个新的变量，将"变量名"设置为 M2_HOME，"变量值"设置为 Maven 的解压目录，如 E:\apache-maven-3.6.1，如图 1-16 所示。

(2) 在"系统变量"列表框中找到名为 Path 的变量，然后单击"编辑"按钮。弹出"编辑环境变量"对话框，在下方输入框中输入"%M2_HOME%\bin"，然后单击"确定"按钮，如图 1-17 所示。

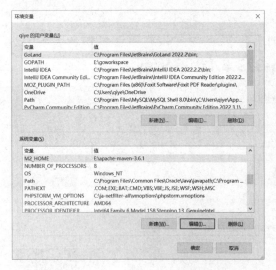

图 1-16 "环境变量"对话框 　　　　　图 1-17 "编辑环境变量"对话框

### 1.2.5 验证安装

配置 Maven 环境变量之后，需要验证配置的正确性，以确保 Maven 可以正常使用。在 CMD 命令行界面输入"mvn -v"并按 Enter 键，如果显示 Maven 的版本信息，则说明安装成功，如图 1-18 所示。

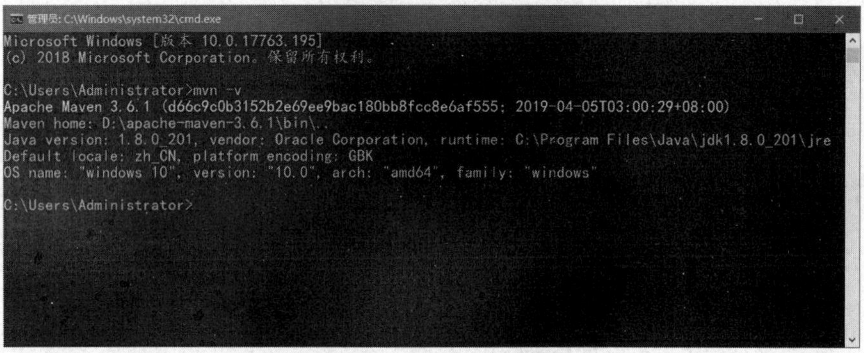

图 1-18 Maven 安装成功

### 1.2.6 修改 Maven 配置文件

当 Java 项目使用 Maven 时，其所需要的第三方依赖需要到 Maven 的网上仓库下载，然后存放到本地。Maven 的默认中央仓库在国外，其下载速度在国内会比较慢。为了加快依赖项的下载速度，在我国国内，一些组织或企业搭建了自己的 Maven 仓库，这些仓库与 Maven 的中央仓库实时同步。国内的开发者可以通过修改 Maven 的配置文件，使 Maven 的默认仓库指向国内的仓库。本书中使用的是阿里云的 Maven 镜像仓库。修改 Maven 配置文件的详细操作步骤如下。

(1) 在 Maven 的安装目录的 conf 目录下(例如 E:\apache-maven-3.6.1\conf)，找到

settings.xml 配置文件，用记事本等工具打开。

(2) 在 E 盘下新建 repository 目录，修改<localRepository>标签为本地仓库地址，修改生效后下载的所有依赖文件将存放在设置的文件夹中。这里把本地仓库指定为本机路径 E:\repository，如图 1-19 所示。

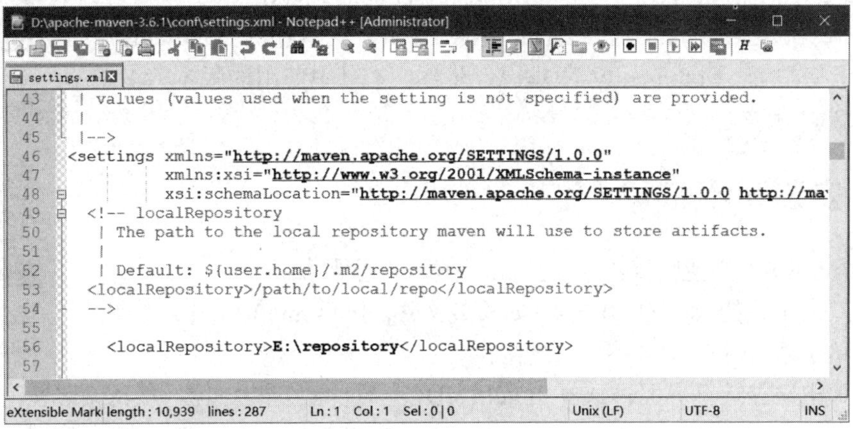

图 1-19 修改 Maven 的配置文件配置本地仓库

(3) 将 Maven 的仓库指向阿里云的镜像仓库，在<mirrors>标签中增加阿里云镜像，如图 1-20 所示。

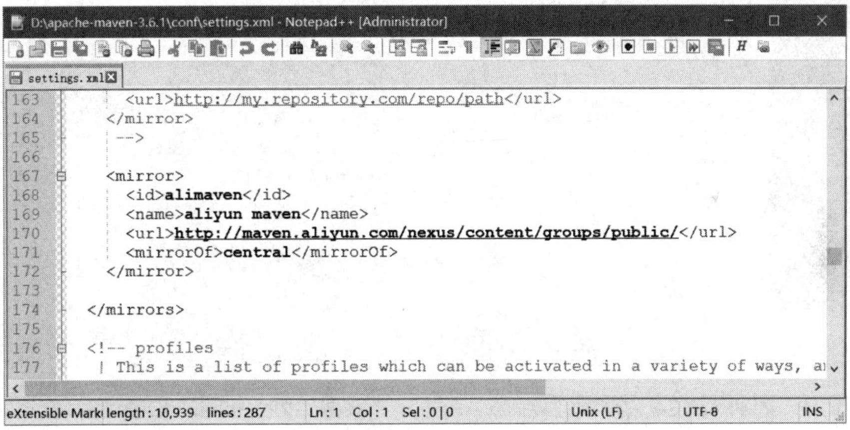

图 1-20 配置阿里云镜像

## 1.3 IntelliJ IDEA 的安装

IntelliJ IDEA 是由 JetBrains 公司开发的一款功能强大的集成开发环境(IDE)，主要用于 Java 语言的开发。它也支持多种其他编程语言，如 Kotlin、Groovy、Scala 等，以及前端开发语言如 JavaScript、TypeScript 和 HTML。IntelliJ IDEA 以其高效的代码自动完成、强大的代码分析、先进的智能重构以及全面的调试工具而闻名，为用户提供了一个功能强大且易于使用的开发环境。

## 1.3.1 下载 IntelliJ IDEA 安装包

打开浏览器，访问 IntelliJ IDEA 官方下载网址：
https://www.jetbrains.com/idea/ download/?section=windows

选择适合自己的版本：有免费的 Community 版本和付费的 Ultimate 版本。对于初学者，Community 版本已经足够。本书中使用的版本为 IDEA 社区版 2023.2 版本。

微课：下载 IntelliJ IDEA

## 1.3.2 安装 IntelliJ IDEA

IntelliJ IDEA 安装包下载完成之后，具体安装步骤如下。

（1）双击下载的 ideaIC-2023.2.exe 安装文件，打开 IntelliJ IDEA 安装欢迎界面，单击 Next 按钮，如图 1-21 所示。

微课：安装 IntelliJ IDEA

（2）弹出路径选择界面，选择 IntelliJ IDEA 软件的安装路径，默认安装在 C:\Program Files\JetBrains\IntelliJ IDEA Community Edition 2023.2 目录下，单击 Next 按钮，如图 1-22 所示。

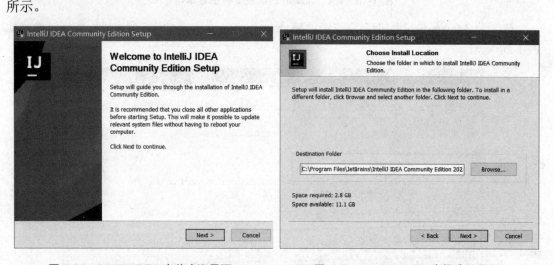

图 1-21　IntelliJ IDEA 安装欢迎界面　　　图 1-22　IntelliJ IDEA 路径选择界面

（3）弹出安装选项界面，选中 Create Desktop Shortcut 选项组中的复选框，表示在桌面创建快捷方式，单击 Next 按钮，如图 1-23 所示。

（4）弹出选择"开始"菜单文件夹界面，单击 Install 按钮，如图 1-24 所示。

（5）弹出安装完成界面，选中 Run IntelliJ IDEA Community Edition 复选框，表示运行 IntelliJ IDEA 软件，单击 Finish 按钮，如图 1-25 所示。

（6）弹出配置 IntelliJ IDEA 界面，选中 Do not import settings 单选按钮，表示不导入设置，单击 OK 按钮，如图 1-26 所示。

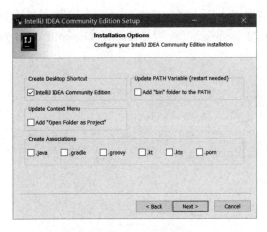

图 1-23　安装选项界面

图 1-24　选择"开始"菜单文件夹界面

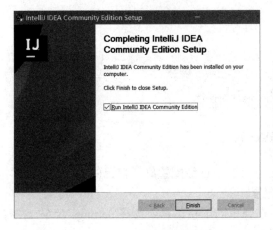

图 1-25　安装完成界面

图 1-26　配置 IntelliJ IDEA 界面

（7）IntelliJ IDEA 安装成功后，单击 New Project 按钮即可创建一个新项目，如图 1-27 所示。

图 1-27　创建新项目

(8) 弹出 New Project 对话框，输入项目名称"chapter01"，将 Location 设置为在计算机中的存储路径，Language 设置为 Java，Build system 设置为 Maven，GroupId 设置为 com.demo。单击 Create 按钮，如图 1-28 所示。

图 1-28  New Project 对话框

## 1.3.3　在 IntelliJ IDEA 中设置 Maven

IDEA 中自带了默认的 Maven 设置，本书在 1.2 节已经配置好了自己的 Maven 工具，因此，此处只需要修改 IDEA 的默认 Maven 配置。

在 IntelliJ IDEA 主界面中，单击 图标，在弹出的下拉菜单中选择 Settings 命令，在打开的 Settings 对话框的左侧找到 Maven 选项并选中，分别将 Maven home path、User settings file、Local repository 设置为我们已经安装好的 Maven 配置路径，如图 1-29 和图 1-30 所示。

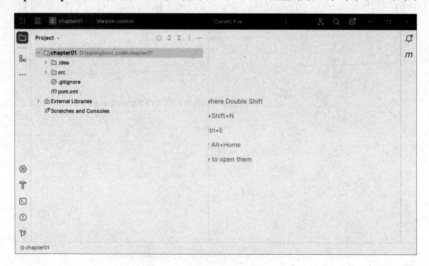

图 1-29  IntelliJ IDEA 主界面

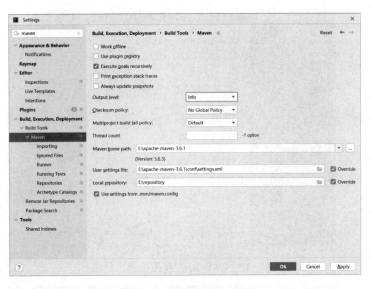

图 1-30 Settings 对话框

设置完毕后单击 OK 按钮即可完成配置。

## 1.4　MySQL 数据库的安装

MySQL 是世界上最受欢迎的开源关系数据库之一。它是一个多线程、多用户的 SQL 数据库管理系统，拥有高效的查询处理性能和可扩展性。MySQL 用途广泛，从嵌入式应用到大型网站、从个人项目到商业软件，都有其身影。

下面我们来介绍 MySQL 的安装。

微课：MySQL 下载

### 1.4.1　MySQL 的安装

（1）本书中使用 MySQL 5.7 版本。首先在 MySQL 的官网 https://dev.mysql.com/downloads/windows/installer/5.7.html 中单击 Download 按钮下载软件，如图 1-31 所示。

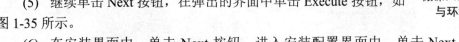

图 1-31　MySQL 下载界面

（2）双击 mysql-installer-community-5.7.43.0.msi 安装包，在弹出的安装类型选择界面中，选中 Custom 单选按钮，如图 1-32 所示。

（3）单击 Next 按钮，进入如图 1-33 所示的选择产品界面。

（4）单击 Next 按钮，进入如图 1-34 所示的安装路径设置界面。

（5）继续单击 Next 按钮，在弹出的界面中单击 Execute 按钮，如图 1-35 所示。

微课：MySQL 的安装与环境变量配置

（6）在安装界面中，单击 Next 按钮，进入安装配置界面中，单击 Next 按钮，如图 1-36 所示。

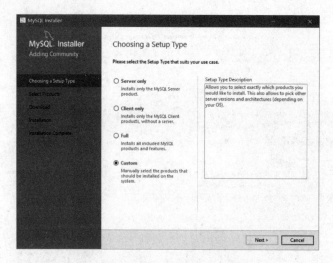

图 1-32　MySQL 安装类型选择界面

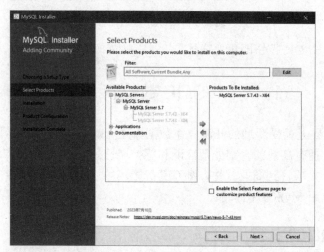

图 1-33　选择产品界面

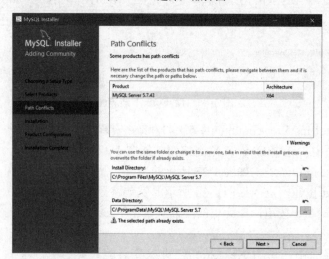

图 1-34　MySQL 安装路径设置界面

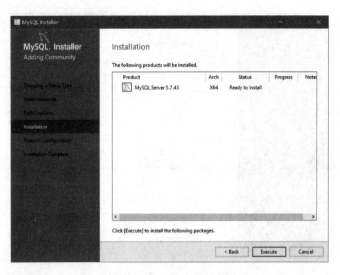

图 1-35　MySQL 开始安装界面

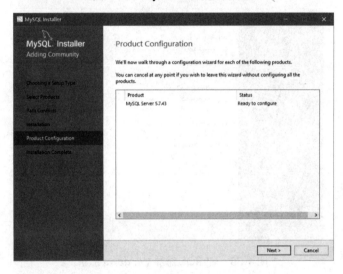

图 1-36　MySQL 安装配置界面

（7）在类型和网络界面中，单击 Next 按钮，在安装过程中需要为 MySQL 的超级管理员 root 账户设置密码，在密码设置界面中，设置"123456"为密码。在密码文本框中输入"123456"，然后在确认密码文本框中再次输入"123456"，最后单击 Next 按钮，如图 1-37 所示。

（8）在添加 Windows 服务界面中，单击 Next 按钮，如图 1-38 所示。

（9）在服务权限设置界面中，单击 Next 按钮，如图 1-39 所示。

（10）在应用配置界面中单击 Execute 按钮，将 Configuration Steps 选项卡中的全部选项均选中，然后单击 Finish 按钮完成安装，如图 1-40 所示。

（11）在计算机桌面上右击"计算机"或"此电脑"图标，在弹出的快捷菜单中选择"属性"命令，打开"系统"窗口，单击"高级系统设置"链接，如图 1-41 所示。

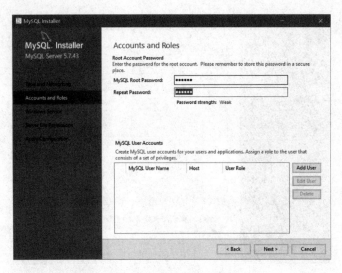

图 1-37　密码设置界面

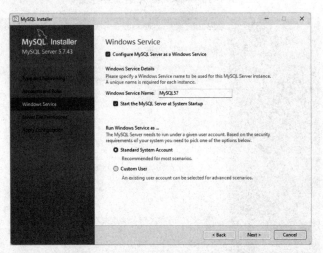

图 1-38　MySQL 添加 Windows 服务

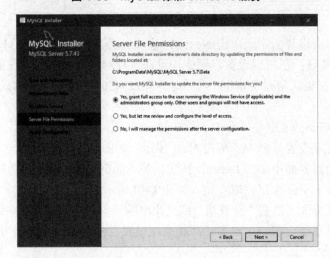

图 1-39　MySQL 服务权限设置界面

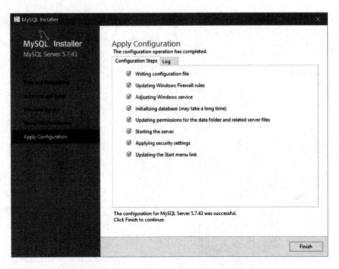

图 1-40 MySQL 安装完成界面

图 1-41 "系统"窗口

(12) 在弹出的"系统属性"对话框中单击"环境变量"按钮,弹出"环境变量"对话框,在"系统变量"列表框中选择 Path 变量,单击"编辑"按钮,如图 1-42、图 1-43 所示。

(13) 在弹出的"编辑环境变量"对话框中,单击"新建"按钮,在下方输入框中输入"C:\Program Files\MySQL\MySQL Server 5.7\bin",单击"确定"按钮,如图 1-44 所示。

(14) 打开 CMD 命令行界面,输入"mysql -uroot -p"命令,密码输入"123456",按 Enter 键,登录 MySQL,如图 1-45 所示。

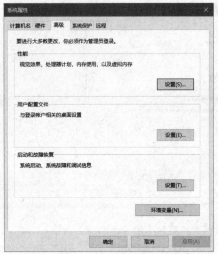

图 1-42 "系统属性"对话框

图 1-43 "环境变量"对话框

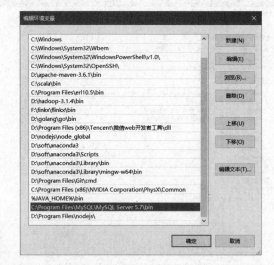

图 1-44 "编辑环境变量"对话框

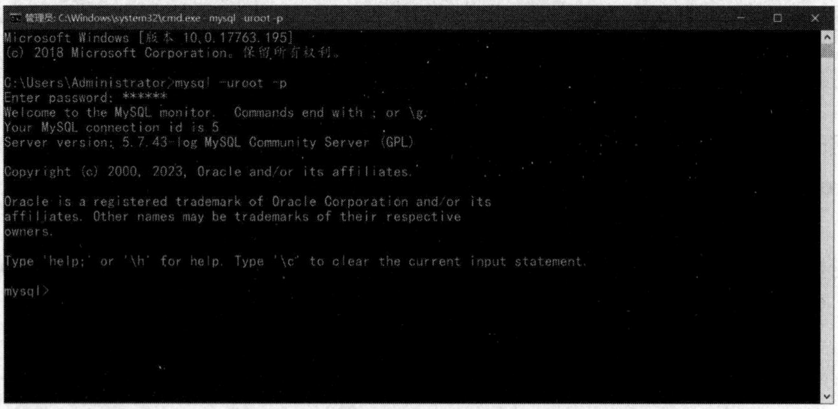

图 1-45 登录 MySQL

## 1.4.2 Navicat 客户端的安装与使用

Navicat 是一款功能强大的数据库管理和开发软件，它允许用户轻松地连接到多种数据库，如 MySQL、MariaDB、MongoDB、SQL Server、Oracle、PostgreSQL 和 SQLite。Navicat 直观的界面设计和强大的功能集合，使得数据库管理和维护更加简单、高效。

微课：Navicat 下载

本书中使用 Navicat 12 版本。具体的安装步骤如下：

（1）从链接 http://download.navicat.com.cn/download/navicat120_premium_cs_x64.exe 下载 Navicat 软件安装包，下载完成后双击安装包，在弹出的欢迎安装界面中单击"下一步"按钮，如图 1-46 所示。

（2）弹出"许可证"界面，在该界面中选中"我同意"单选按钮，然后单击"下一步"按钮。

微课：Navicat 软件安装、创建数据库和表

（3）弹出"选择安装文件夹"界面，在其中设置 Navicat 的安装路径，单击"下一步"按钮。

（4）弹出"选择 开始 目录"界面，在其中设置创建快捷方式的目录，单击"下一步"按钮，如图 1-47 所示。

图 1-46  Navicat 欢迎安装界面　　　　图 1-47  "选择 开始 目录"界面

（5）弹出"选择额外任务"界面，选中 Create a desktop icon 复选框，然后单击"下一步"按钮，如图 1-48 所示。

（6）弹出"准备安装"界面，单击"安装"按钮，如图 1-49 所示。

（7）安装成功之后，可在计算机桌面上找到 Navicat 的快捷方式。双击快捷方式图标即可启动 Navicat。

（8）在 Navicat Premium 窗口中，单击左上角的"连接"按钮，在弹出的下拉菜单中选择 MySQL 命令，如图 1-50 所示。

（9）在弹出的"MySQL-新建连接"对话框中，设置"连接名"为 localhost，设置"密码"为 123456，单击"连接测试"按钮，如图 1-51 所示。

（10）弹出连接成功对话框，单击"确定"按钮，如图 1-52 所示。

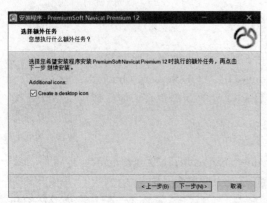
图 1-48 "选择额外任务"界面

图 1-49 "准备安装"界面

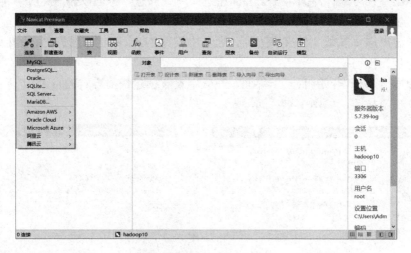
图 1-50 选择 MySQL 命令

图 1-51 "MySQL-新建连接"对话框

图 1-52 连接 MySQL 测试成功

(11) 在 Navicat Premium 主界面中双击 localhost 选项,可以显示其下的系统数据库,如图 1-53 所示。

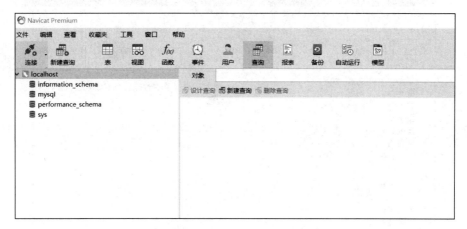

图 1-53　Navicat Premium 主界面

### 1.4.3　MySQL 和 Navicat 的简单使用

本小节简单介绍如何使用 Navicat 工具操作和管理 MySQL 数据库。主要包含创建测试数据库，在创建的数据库中创建数据表，以及在表中添加数据和进行数据查询等。其详细步骤如下。

（1）在 MySQL 中创建一个数据库。选中 localhost 选项，右击，在弹出的快捷菜单中选择"新建数据库"命令，如图 1-54 所示。

（2）弹出"新建数据库"对话框，在其中设置"数据库名"为 test1，"字符集"为 utf8，然后单击"确定"按钮，如图 1-55 所示。

图 1-54　选择"新建数据库"命令

图 1-55　"新建数据库"对话框

（3）在 Navicat Premium 主界面中双击 test1 数据库，选中"查询"选项，单击"新建查询"按钮，打开 SQL 编辑器窗口，如图 1-56、图 1-57 所示。

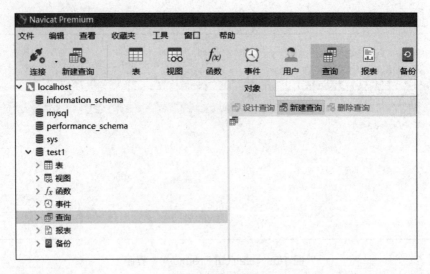

图 1-56 新建查询

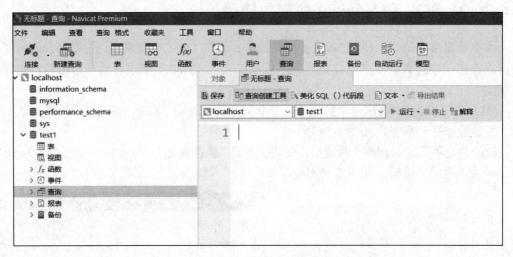

图 1-57 SQL 编辑器窗口

(4) 在 SQL 编辑器窗口中，创建一张用户表 t_user。

创建数据表的 SQL 语法如下：

```
create table 表名(
    字段名1  数据类型 [约束],
    字段名2  数据类型 [约束]
)
```

例如，创建一个用户表，包含标识(id)、姓名(name)和年龄(age)三个字段，SQL 语句如下：

```
create table t_user(
    id   int  primary key auto_increment,
    name varchar(50),
    age  int
)
```

其中，int 是数值类型，varchar 是字符串类型。primary key 代表主键约束，该列值不能重复且不为空。auto_increment 可以实现主键自增长。

在 Navicat 中执行以上 SQL 语句，结果如图 1-58 所示。

图 1-58　执行建表语句

（5）向表中添加数据。选中添加语句，然后单击"运行已选择的"按钮，如图 1-59 所示。

（6）查询表中的数据，如图 1-60 所示。

图 1-59　向表中添加数据　　　　　图 1-60　查询语句执行结果

## 1.5　第一个 SpringBoot 程序

通过前面介绍的内容，我们已经完成了所有的开发环境的准备，下面将基于以上的准备工作，开发一个简单的 SpringBoot 程序。

在 1.3 节的内容中，已经在 IntelliJ IDEA 中创建了 Maven 项目 chapter01，下面将继续在这个项目中完成第一个 SpringBoot 项目的开发。

本节先完成一个简单项目的搭建和运行，代码中的一些细节将不做详细解释，在后续章节中，会详细讲解其具体的作用和用法。

动画：SpringBoot 在现代软件开发中的应用与优势

### 1.5.1　添加依赖

在 pom.xml 文件中添加 SpringBoot 项目需要的 Maven 依赖，添加依赖后 pom.xml 的完整代码如下：

微课：SpringBoot 初体验

```
<?xml version="1.0" encoding="UTF-8"?>
```

```xml
<project xmlns="http://maven.apache.org/POM/4.0.0"
    xmlns:xsi="http://www.w3.org/2001/XMLSchema-instance"
    xsi:schemaLocation="http://maven.apache.org/POM/4.0.0
        http://maven.apache.org/xsd/maven-4.0.0.xsd">
    <modelVersion>4.0.0</modelVersion>

    <groupId>com.demo</groupId>
    <artifactId>chapter01</artifactId>
    <version>1.0-SNAPSHOT</version>

    <properties>
        <maven.compiler.source>8</maven.compiler.source>
        <maven.compiler.target>8</maven.compiler.target>
        <project.build.sourceEncoding>UTF-8</project.build.sourceEncoding>
    </properties>
    <!--设置Maven项目的parent,此配置可以引入SpringBoot的所有核心依赖-->
    <parent>
        <groupId>org.springframework.boot</groupId>
        <artifactId>spring-boot-starter-parent</artifactId>
        <version>2.4.4</version>
    </parent>

    <dependencies>
        <!--引入SpringBoot中开发Web服务的依赖 -->
        <dependency>
            <groupId>org.springframework.boot</groupId>
            <artifactId>spring-boot-starter-web</artifactId>
        </dependency>
    </dependencies>
</project>
```

依赖添加完成之后,单击 IDEA 右侧选项卡 Maven 中的更新按钮下载依赖,如图 1-61 所示。

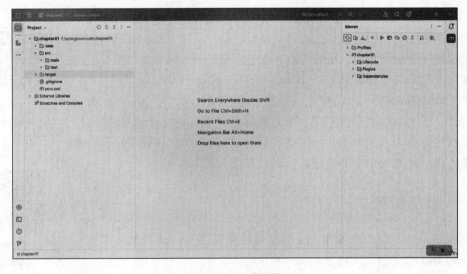

图 1-61　下载依赖

## 1.5.2 创建启动类和控制器

创建启动类和控制器的步骤如下。

(1) 使用 IDEA 在项目 chapter01 的 src/main/java 目录下创建一个 Java 包,命名为 com.demo,如图 1-62、图 1-63 所示。

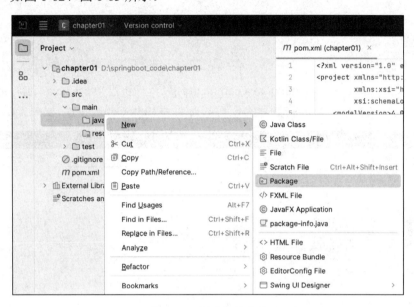

图 1-62 创建包的菜单

图 1-63 创建包

(2) 在 com.demo 包下创建 SpringBoot 程序启动类,将其命名为 MySpringBoot,如图 1-64 所示。

图 1-64 创建启动类

MySpringBoot 类的代码如下:

```
package com.demo;

import org.springframework.boot.SpringApplication;
import org.springframework.boot.autoconfigure.SpringBootApplication;
@SpringBootApplication              //标记MySpringBoot为主程序启动类
public class MySpringBoot {

    public static void main(String[] args) {
        SpringApplication.run(MySpringBoot.class);
    }
}
```

(3) 在 chapter01 的 src/main/java/com.demo 目录下创建一个 Java 包,将其命名为 controller,并且创建一个 Java 控制器类,将其命名为 HelloController,如图 1-65 所示。

图 1-65 创建控制器

HelloController 类完整代码如下:

```
package com.demo.controller;

import org.springframework.stereotype.Controller;
import org.springframework.web.bind.annotation.RequestMapping;
import org.springframework.web.bind.annotation.ResponseBody;

@Controller
public class HelloController {

    @RequestMapping("/hello")
    @ResponseBody
    public String hello(){
        return "hello SpringBoot";
    }
}
```

(4) 在启动类 MySpringBoot.java 选项卡中单击左侧的三角箭头▷,然后选择 Run 'MySpringBoot.main()'运行启动类,如图 1-66、图 1-67 所示。

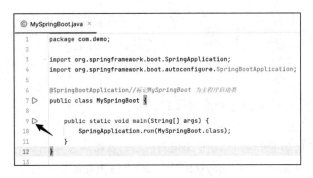
图 1-66 MySpringBoot.java 选项卡

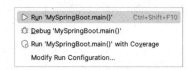
图 1-67 运行启动类

### 1.5.3 测试

在浏览器中输入项目的访问地址 http://localhost:8080/hello，结果如图 1-68 所示。

图 1-68 访问项目

## 本章小结

本章介绍了 SpringBoot 开发环境的准备过程。为了确保读者能够顺利进行后续的开发，本章首先明确了学习目标与学习要点，让每位读者都能了解学习的重点和难点。

本章重点介绍了 JDK 的重要性，并提供了详细的安装和配置教程，使读者能够为 Java 开发做好准备。接着，详细讲解了 IntelliJ IDEA，一款功能非常强大的 Java 开发工具的安装与配置过程，通过对其基本操作的介绍，读者能够利用这个工具更高效地进行开发。Maven，作为一款功能非常强大的项目管理和自动化构建工具，我们也进行了详细的讲解。在数据库方面，我们选择 MySQL 作为学习的重点，为读者提供了完整的安装和配置教程，以满足后续开发中数据存储的需求。最后介绍了 SpringBoot 的项目结构和基本配置，为后续的项目开发打下坚实的基础。

总的来说，本章为读者提供了一套完整的 SpringBoot 开发工具链的安装和配置教程。在掌握了这些工具后，读者就能为后续的项目开发做好充分准备，希望每位读者都能够在本章的学习中收获满满。此外，本章只是使用 SpringBoot 框架创建了一个最初级的应用，而要开发一个实际的项目还会用到 MyBatis、Spring、SpringMVC 等框架，本书在接下来的章节中会先介绍各个框架的使用，在最后一章中会详细介绍使用上述 4 个框架集成开发一个功能完备的系统。

## 课 后 习 题

一、选择题

1. 在 SpringBoot 中，（　　）工具用于项目构建和依赖管理。
   A. Eclipse  　　B. IntelliJ IDEA  　　C. Maven  　　D. Gradle
2. 为了在 SpringBoot 应用程序中使用数据库，通常会使用（　　）数据库管理系统。
   A. MongoDB  　　B. PostgreSQL  　　C. MySQL  　　D. Redis
3. 要想在 SpringBoot 应用程序中使用 Maven，我们需要在项目的（　　）文件中定义依赖关系。
   A. application.properties  　　B. pom.xml
   C. application.yml  　　D. build.gradle

二、填空题

1. SpringBoot 的主要目标是简化 ＿＿＿＿＿＿ 应用程序的开发。
2. SpringBoot 应用程序的入口点通常是一个包含 public static void main(String[] args) 方法的 Java 类，这个类被称为 ＿＿＿＿＿＿。

三、判断题

1. SpringBoot 是一个基于 Spring Framework 的独立项目，用于简化 Spring 应用程序的开发。　　（　　）
2. SpringBoot 应用程序的入口点通常是一个包含 main 方法的 Java 类。　　（　　）

四、简答题

1. 什么是 Apache Maven？简要说明其在 Java 项目管理中的作用。
2. MySQL 是一种什么类型的数据库？简要说明其主要特点。

五、实操题

1. 请按照教程，安装 JDK、Maven、MySQL，并配置相应的环境变量。
2. 在 IntelliJ IDEA 中创建一个简单的 SpringBoot 项目，并使其运行，显示"Hello, SpringBoot!"。

### 软件是构建现代社会的基石

在准备 SpringBoot 开发环境时，我们应深刻认识到软件开发的重要性。软件不仅是构建现代社会的基础，而且是实现信息化、智能化的关键手段。因此，作为未来的软件工程师，我们不仅要懂得如何搭建高效的开发环境，更需要明白自己在社会发展中所承担的角色和责任。

# 第 2 章
# MyBatis 框架初体验

### 学习目标

1. 了解 MyBatis 是什么，并理解它在 Java 世界中为何如此受欢迎。

2. 能够设置和使用 MyBatis 的基础配置，包括与数据库的连接、事务管理等。

3. 熟悉基础的 CRUD 操作：理解如何使用 MyBatis 进行基本的数据操作，例如插入、查询、更新和删除。

### 学习要点

1. 简单了解 MyBatis 的历史。

2. 掌握 MyBatis 开发环境的搭建，以及配置文件的编写，并完成 CRUD 操作。

本章知识点结构如图 2-1 所示。

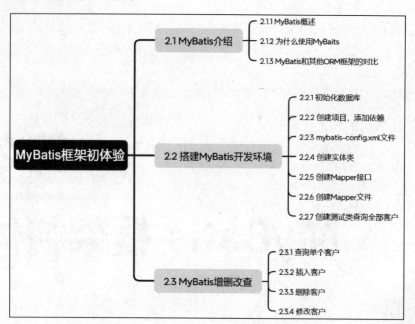

图 2-1　MyBatis 框架初体验

## 2.1　MyBatis 介绍

动画：MyBatis 在数据库交互中的作用与优势

### 2.1.1　MyBatis 概述

MyBatis 是一个优秀的持久层框架，它充当了 Java 应用程序与数据库之间的桥梁，使得开发者可以用更少的时间和精力去处理数据库操作的复杂性，同时又保留了对 SQL 的完全控制。MyBatis 消除了几乎所有的 JDBC 代码和参数的手动设置以及结果集的检索，使得数据库代码更简洁，更易于维护。

MyBatis 起源于 2002 年，最初被称为 iBatis，由 Clinton Begin 开发并被 Apache Software Foundation 接管。iBatis 是一种 SQL 映射器，用于将 SQL 查询结果映射到对象模型。在 2010 年，iBatis 被转移到 Google Code，并更名为 MyBatis。2013 年，MyBatis 从 Google Code 迁移到了 Github。在这段时间里，MyBatis 的使用者社区不断壮大，并且在全球范围内的许多企业和项目中得到了广泛的应用。

### 2.1.2　为什么使用 MyBatis

MyBatis 在数据库操作领域的独特性，使得它在许多场景中成为首选的持久层框架，MyBatis 具有以下特点。

**1. 简单易用**

MyBatis 的学习成本相对较低，对于开发者来说，使用它可以节省大量的开发时间和精力。开发者只需要定义 SQL 语句和映射规则，就可以实现复杂的数据库操作。

### 2. 灵活

MyBatis 没有强制的数据模型规范，开发者可以自由地定义 SQL 语句和映射规则，这给了开发者极大的灵活性。对于复杂的数据库操作，使用 MyBatis 可以编写优化的 SQL，实现更高的运行效率。

### 3. 易于集成

MyBatis 提供了多种集成方式，可以很容易地与 Spring、SpringBoot 等其他框架集成。对于 Java 企业级应用来说，这无疑是一个巨大的优点。

## 2.1.3 MyBatis 和其他 ORM 框架的对比

在 Java 的持久化框架领域，MyBatis 与 Hibernate、JPA 等是常见的比较对象。这些框架各有优点，具体选用哪一个，取决于项目的需求。MyBatis 的主要特点是自由和灵活，用户可以自己编写 SQL，决定如何映射结果集；尽管 Hibernate、JPA 等提供了数据操作的全自动化处理，对于不太复杂的应用则可以提高开发效率。然而，当遇到复杂映射和性能要求较高的场景时，Hibernate 等全自动化框架可能力不从心，而 MyBatis 的灵活性却能够更好地满足这些需求。

## 2.2 搭建 MyBatis 开发环境

本节将通过一个简单的客户信息管理系统案例，带领读者学习 MyBatis 的基本使用。这是一个初级的客户信息管理系统，通过该系统可以帮助企业管理客户信息，其核心功能包括添加新的客户信息、修改客户信息、删除客户信息、查询客户信息等。

微课：MyBatis 开发环境搭建

### 2.2.1 初始化数据库

首先通过 MySQL 客户端创建客户信息数据表，并且初始化一些测试数据。创建数据库和数据表的 SQL 语句如下：

```sql
# 创建一个新的数据库
CREATE DATABASE test_mybatis DEFAULT CHARACTER SET utf8;

# 切换到 test_mybatis 数据库下
use test_mybatis;

# 创建一张客户表
create table t_customer(
    id       INT PRIMARY KEY AUTO_INCREMENT,
    username VARCHAR(20) NOT NULL,
    city     VARCHAR(50),
    birthday DATE
);
```

数据库和数据表创建完成后，可以向数据表中添加几条测试数据，SQL 语句如下：

```
insert into t_customer values(null,'张三','北京','2001-10-08');
insert into t_customer values(null,'李四','北京','2002-02-01');
insert into t_customer values(null,'张丽','上海','2001-07-12');
```

在 MySQL 命令行客户端执行上述代码后,可以查看到如图 2-2 所示的结果。

图 2-2  查询结果

### 2.2.2  创建项目,添加依赖

准备好数据库之后,下面将使用 IDEA 创建一个 Maven 项目,为本章后续 MyBatis 的操作做准备。创建 Maven 项目的详细过程如下。

(1) 通过 IDEA 创建一个项目,设置项目名称 Name 为 chapter02。Location 代表项目的本地磁盘存储路径,设置 Language 为 Java,将 Build system 设置为 Maven,取消选中 Add sample code 复选框,设置 GroupId 为 com.demo,如图 2-3 所示。

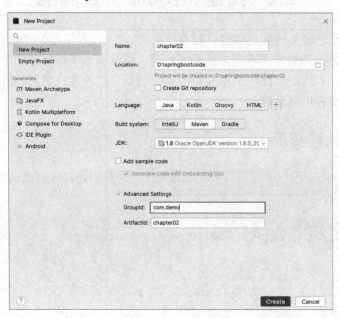

图 2-3  创建项目

(2) 在新创建的 Maven 项目下找到 pom.xml 文件,在<dependencies>标签中添加相应的依赖,代码如下:

```
<dependencies>
    <!-- MyBatis 的依赖 -->
```

```xml
    <dependency>
        <groupId>org.mybatis</groupId>
        <artifactId>mybatis</artifactId>
        <version>3.5.6</version>
    </dependency>

    <!-- mySQL 的依赖 -->
<dependency>
        <groupId>mysql</groupId>
        <artifactId>mysql-connector-java</artifactId>
        <version>5.1.49</version>
    </dependency>

    <!-- 单元测试的依赖 -->
    <dependency>
        <groupId>junit</groupId>
        <artifactId>junit</artifactId>
        <version>4.12</version>
        <scope>test</scope>
    </dependency>
</dependencies>
```

上述依赖中：

```xml
<groupId>org.mybatis</groupId>
<artifactId>mybatis</artifactId>
<version>3.5.6</version>
```

为开发 MyBatis 程序所需要引入的 MyBatis 官方依赖。本书将一直使用 3.5.6 版本。

```xml
<groupId>mysql</groupId>
<artifactId>mysql-connector-java</artifactId>
<version>5.1.49</version>
```

为 Java 连接和使用 MySQL 数据库所需要的依赖。本书将一直使用 5.1.49 版本。

```xml
<groupId>junit</groupId>
<artifactId>junit</artifactId>
<version>4.12</version>
<scope>test</scope>
```

为编写单元测试使用的工具 junit 的依赖。本书将一直使用 4.12 版本。

### 2.2.3　mybatis-config.xml 文件

在项目的 resources 目录下创建一个 mybatis-config.xml 文件，这是 MyBatis 的核心配置文件，一般需要配置数据库的连接信息，还有后续的 Mapper 文件的路径等，代码如下：

```xml
<?xml version="1.0" encoding="UTF-8" ?>
<!DOCTYPE configuration PUBLIC "-//mybatis.org//DTD Config 3.0//EN"
        "http://mybatis.org/dtd/mybatis-3-config.dtd">
<configuration>
```

```xml
<!-- 打印日志 -->
<settings>
    <setting name="logImpl" value="STDOUT_LOGGING"/>
</settings>

<!-- 连接数据库信息 -->
<environments default="development">
    <environment id="development">
        <transactionManager type="JDBC"/>
        <dataSource type="POOLED">
            <!-- MySQL 驱动 -->
            <property name="driver" value="com.mysql.jdbc.Driver"/>
            <!-- MySQL 数据库地址 -->
            <property name="url"
                value="jdbc:mysql://localhost:3306/test_mybatis?
                characterEncoding=utf8"/>
            <!-- MySQL 数据库用户名 -->
            <property name="username" value="root"/>
            <!-- MySQL 数据库密码 -->
            <property name="password" value="123456"/>
        </dataSource>
    </environment>
</environments>

</configuration>
```

### 2.2.4 创建实体类

在项目的 src/main/java 目录下创建 com.demo.pojo 包，新建一个 Customer 实体类。这个实体类中的属性和前边的 t_customer 表中的字段一一对应。Customer 实体类的用途是存储数据库表中查询的结果以及用户要添加的数据等。代码如下：

```java
package com.demo.pojo;

public class Customer {

    private Integer  id;
    private String   username;
    private String   city;
    private Date     birthday;

    //生成上面属性的 get、set 方法以及 toString 方法
}
```

### 2.2.5 创建 Mapper 接口

在项目的 src/main/java 下创建 com.demo.mapper 包，新建一个 CustomerMapper 接口文件，声明查询全部客户的方法。该方法的返回值是 List 集合，用于存储数据库客户表中的

查询结果。代码如下:

```
package com.demo.mapper;
import com.demo.pojo.Customer;
import java.util.List;
public interface CustomerMapper {
    public List<Customer> queryAllCustomer();
}
```

### 2.2.6 创建 Mapper 文件

在 resources 目录下创建 mapper 目录,在该目录下创建一个名为 CustomerMapper.xml 的配置文件,该文件用于配置要执行的 SQL 语句。<mapper>标签中的 namespace 属性应配置为接口的全类名,<select>标签的 id 属性应配置为接口中定义的方法名。在<select>标签中编写用于检索全部客户的 SQL 查询语句,resultType 属性应配置为实体类的全类名。代码如下:

```xml
<?xml version="1.0" encoding="UTF-8"?>
<!DOCTYPE mapper
  PUBLIC "-//mybatis.org//DTD Mapper 3.0//EN"
  "http://mybatis.org/dtd/mybatis-3-mapper.dtd">

<mapper namespace="com.demo.mapper.CustomerMapper">

    <select id="queryAllCustomer" resultType="com.demo.pojo.Customer">
        select * from t_customer
    </select>

</mapper>
```

CustomerMapper.xml 需要在 mybatis-config.xml 中注册。在 mybatis-config.xml 文件的<Configuration>标签中添加以下代码:

```xml
<mappers>
    <mapper resource="mappers/CustomerMapper.xml"/>
</mappers>
```

### 2.2.7 创建测试类查询全部客户

以上完成了项目的搭建和代码开发。为了验证所有步骤和程序的正确性,下面编写一个测试程序,来测试 MyBatis 程序是否能正确运行。

在项目 chapter02 的 src/test/java 目录下创建 Java 包 com.demo,在该包下创建 TestMyBatis 测试类,代码如下:

```java
package com.demo;

import com.demo.mapper.CustomerMapper;
import com.demo.pojo.Customer;
import org.apache.ibatis.io.Resources;
import org.apache.ibatis.session.SqlSession;
import org.apache.ibatis.session.SqlSessionFactory;
import org.apache.ibatis.session.SqlSessionFactoryBuilder;
import org.junit.Test;

import java.io.IOException;
import java.io.InputStream;
import java.util.List;

public class TestMyBatis {

    @Test
    public void testQueryAll() throws IOException {
        //1.加载配置文件信息
        String resource = "mybatis-config.xml";
        InputStream inputStream = Resources.getResourceAsStream(resource);
        //2.创建一个连接池
        SqlSessionFactory sqlSessionFactory =
            new SqlSessionFactoryBuilder().build(inputStream);
        //3.获取一个连接
        SqlSession session = sqlSessionFactory.openSession();
        //4.获取执行SQL的代理对象
        CustomerMapper customerMapper =
             session.getMapper(CustomerMapper.class);
        //5.执行SQL
        List<Customer> lists = customerMapper.queryAllCustomer();
        //6.处理结果
        for (Customer customer : lists) {
            System.out.println(customer);
        }
        //7.关闭资源
        session.close();
    }
}
```

测试执行结果如图2-4所示，数据库表中的所有客户信息均会被查询出来。表明以上步骤中的环境搭建内容均是正确的。

```
Created connection 1438030319.
Setting autocommit to false on JDBC Connection [com.mysql.jdbc.JDBC4Connection@55b699ef]
==>  Preparing: select * from t_customer
==> Parameters:
<==    Columns: id, username, city, birthday
<==        Row: 1, 张三, 北京, 2001-10-08
<==        Row: 2, 李四, 北京, 2002-02-01
<==        Row: 3, 张丽, 上海, 2001-07-12
<==      Total: 3
Person{id=1, username='张三', city='北京', birthday='Mon Oct 08 00:00:00 CST 2001'}
Person{id=2, username='李四', city='北京', birthday='Fri Feb 01 00:00:00 CST 2002'}
Person{id=3, username='张丽', city='上海', birthday='Thu Jul 12 00:00:00 CST 2001'}
Resetting autocommit to true on JDBC Connection [com.mysql.jdbc.JDBC4Connection@55b699ef]
Closing JDBC Connection [com.mysql.jdbc.JDBC4Connection@55b699ef]
Returned connection 1438030319 to pool.
```

图2-4  查询所有客户信息

## 2.3 MyBatis 增删改查

2.2 节的内容主要完成了一个 MyBatis 项目的环境搭建和简单的使用测试。在当前的开发场景中，对数据库数据的操作主要包括查询数据、插入数据、删除数据、修改数据等。本节将在 2.2 节的基础上介绍使用 MyBatis 完成各类数据操作的编写和测试过程。

### 2.3.1 查询单个客户

在真实的开发场景中，最常用的查询场景就是使用数据表的主键为条件查询此条数据。这里的代码使用客户编号来查询对应的客户信息。详细的开发步骤如下。

微课：查询单个客户信息

(1) 在 CustomerMapper 接口中先声明一个方法，参数是客户编号，返回值是客户对象，代码如下：

```
public Customer queryOneCustomer(int id);
```

(2) 在 CustomerMapper.xml 文件中添加一个<select>标签。代码如下：

```xml
<select id="queryOneCustomer" resultType="com.demo.pojo.Customer">
    select * from t_customer where id = #{id}
</select>
```

上述代码中<select>标签的 id 是此标签的标识符，在当前 XML 文件中需唯一，其值 queryOneCustomer 要和 CustomerMapper 中对应的方法名一致。resultType 是查询的结果要映射的 Java 类型，这里是 Customer 类的完整路径。<select>标签内的 SQL 语句是此查询运行时要在数据库中执行的 SQL 语句。其中，#{id}表示此查询会接收一个 Java 程序传过来的参数，参数名为 id，与 CustomerMapper 接口中 queryOneCustomer(int id)函数的参数 id 相对应。

(3) 在测试类 TestMyBatis 中增加 testQueryOne()方法，该方法用于查询客户编号为 1 的客户信息。代码如下：

```java
@Test
public void testQueryOne() throws IOException {
    //1.加载配置文件信息
    String resource = "mybatis-config.xml";
    InputStream inputStream = Resources.getResourceAsStream(resource);
    //2.创建一个连接池
    SqlSessionFactory sqlSessionFactory =
        new SqlSessionFactoryBuilder().build(inputStream);
    //3.获取一个连接
    SqlSession session = sqlSessionFactory.openSession();
    //4.获取执行 SQL 的代理对象
    CustomerMapper customerMapper =
        session.getMapper(CustomerMapper.class);
    //5.执行 SQL
    Customer customer = customerMapper.queryOneCustomer(1);
    //6.处理结果
    System.out.println(customer);
```

```
//7.关闭资源
    session.close();
}
```

测试执行结果如图 2-5 所示,结果只显示 id 为 1 的客户对应的信息。

```
Created connection 825936265.
Setting autocommit to false on JDBC Connection [com.mysql.jdbc.JDBC4Connection@313ac989]
==>  Preparing: select * from t_customer where id = ?
==> Parameters: 1(Integer)
<==    Columns: id, username, city, birthday
<==        Row: 1, 张三, 北京, 2001-10-08
<==      Total: 1
Person{id=1, username='张三', city='北京', birthday='Mon Oct 08 00:00:00 CST 2001'}
Resetting autocommit to true on JDBC Connection [com.mysql.jdbc.JDBC4Connection@313ac989]
Closing JDBC Connection [com.mysql.jdbc.JDBC4Connection@313ac989]
Returned connection 825936265 to pool.
```

图 2-5 查询单个客户信息的结果

微课:
插入客户信息

## 2.3.2 插入客户

下面将介绍使用 MyBatis 向客户表中插入数据的程序开发和测试步骤。

(1) 在 CustomerMapper 接口中声明一个 saveCustomer()方法,参数是客户对象,返回值为 int,代表对数据库影响的行数。代码如下:

```java
public int saveCustomer(Customer customer);
```

(2) 在 CustomerMapper.xml 文件中插入一个<insert>标签,用于存储要执行的插入语句,<insert>标签的 id 属性名应该和接口中的方法名保持一致,#{username}、#{city}、#{birthday}中的名称要和 Customer 中的属性名保持一致,目的是获取传递过来的客户信息数据。代码如下:

```xml
<insert id="saveCustomer">
    insert into t_customer values(null,#{username},#{city},#{birthday})
</insert>
```

(3) 在测试类 TestMyBatis 中添加一个 testSave()方法,在该方法中创建一个客户对象,并且为客户的姓名等属性赋值,然后将该对象中的数据存储到数据库中。代码如下:

```java
@Test
public void testSave() throws IOException, ParseException {
    //1.加载配置文件信息
    String resource = "mybatis-config.xml";
    InputStream inputStream = Resources.getResourceAsStream(resource);
    //2.创建一个连接池
    SqlSessionFactory sqlSessionFactory =
        new SqlSessionFactoryBuilder().build(inputStream);
    //3.获取一个连接
    SqlSession session = sqlSessionFactory.openSession();
    //4.获取执行 SQL 的代理对象
    CustomerMapper customerMapper =
        session.getMapper(CustomerMapper.class);
    //5.执行 SQL
```

```
Customer customer = new Customer();
customer.setUsername("赵六");
customer.setCity("重庆");

SimpleDateFormat sdf = new SimpleDateFormat("yyyy-MM-dd");
Date date = sdf.parse("1998-10-02");
customer.setBirthday(date);
int result = customerMapper.saveCustomer(customer);
//6.处理结果
if(result > 0){
    System.out.println("插入成功");
}else{
    System.out.println("插入失败");
}
//7.提交事务
session.commit();
//8.关闭资源
session.close();
```

测试执行结果如图 2-6 所示，客户信息保存成功。

```
Created connection 403147759.
Setting autocommit to false on JDBC Connection [com.mysql.jdbc.JDBC4Connection@18078bef]
==>  Preparing: insert into t_customer values(null,?,?,?)
==>  Parameters: 赵六(String), 重庆(String), 1998-10-02 00:00:00.0(Timestamp)
<==    Updates: 1
添加成功
Committing JDBC Connection [com.mysql.jdbc.JDBC4Connection@18078bef]
Resetting autocommit to true on JDBC Connection [com.mysql.jdbc.JDBC4Connection@18078bef]
Closing JDBC Connection [com.mysql.jdbc.JDBC4Connection@18078bef]
Returned connection 403147759 to pool.
```

图 2-6  保存客户信息的结果

### 2.3.3  删除客户

下面将介绍使用 MyBatis 从客户表中删除数据的程序开发和测试步骤。

（1）在 CustomerMapper 接口中声明一个 removeCustomerById()方法，参数是 int，用于存储要删除客户的编号，返回值是 int，代表影响数据库的行数。代码如下：

```
public int removeCustomerById(int id);
```

（2）在 CustomerMapper.xml 文件中添加一个<delete>标签，该标签用于编写删除数据的 SQL 语句，其中#{id}用于获取传递过来要删除的客户编号。代码如下：

```
<delete id="removeCustomerById">
    delete from t_customer where id = #{id}
</delete>
```

（3）在测试类 TestMyBatis 中添加一个 testRemove()方法，其作用是将客户编号为 1 的信息从数据库中删除。代码如下：

```java
@Test
public void testRemove() throws IOException, ParseException {
    //1.加载配置文件信息
    String resource = "mybatis-config.xml";
    InputStream inputStream = Resources.getResourceAsStream(resource);
    //2.创建一个连接池
    SqlSessionFactory sqlSessionFactory =
    new SqlSessionFactoryBuilder().build(inputStream);
    //3.获取一个连接
    SqlSession session = sqlSessionFactory.openSession();
    //4.获取执行SQL的代理对象
    CustomerMapper customerMapper =
    session.getMapper(CustomerMapper.class);
    //5.执行SQL
    int result = customerMapper.removeCustomerById(1);
    //6.处理结果
    if(result > 0){
        System.out.println("删除成功");
    }else{
        System.out.println("删除失败");
    }
    //7.提交事务
    session.commit();
    //8.关闭资源
    session.close();
}
```

测试执行结果如图 2-7 所示，客户信息删除成功。

```
Created connection 1438030319.
Setting autocommit to false on JDBC Connection [com.mysql.jdbc.JDBC4Connection@55b699ef]
==>  Preparing: delete from t_customer where id = ?
==> Parameters: 1(Integer)
<==    Updates: 1
删除成功
Committing JDBC Connection [com.mysql.jdbc.JDBC4Connection@55b699ef]
Resetting autocommit to true on JDBC Connection [com.mysql.jdbc.JDBC4Connection@55b699ef]
Closing JDBC Connection [com.mysql.jdbc.JDBC4Connection@55b699ef]
Returned connection 1438030319 to pool.
```

图 2-7  删除客户信息的结果

### 2.3.4  修改客户

下面将介绍使用 MyBatis 修改客户表中数据的程序开发和测试步骤。

(1) 在 CustomerMapper 接口中声明一个 modifyCustomer()方法，参数是客户对象，用于存储要修改客户的信息，返回值是 int，代表影响数据库的行数。代码如下：

```java
public int modifyCustomer(Customer customer);
```

(2) 在 CustomerMapper.xml 文件中添加一个<update>标签，该标签用于编写修改数据的 SQL 语句，其中#{id}用于获取传递过来要修改客户的编号。代码如下：

```xml
<update id="modifyCustomer">
    update  t_customer set username = #{username},city = #{city},
    birthday = #{birthday}   where id = #{id}
</update>
```

(3) 在测试类 TestMyBatis 中添加一个 testModify()方法，其作用是对客户编号为 2 的姓名、城市、生日等信息进行修改。代码如下：

```java
@Test
public void testModify() throws IOException, ParseException {
    //1.加载配置文件信息
    String resource = "mybatis-config.xml";
    InputStream inputStream = Resources.getResourceAsStream(resource);
    //2.创建一个连接池
    SqlSessionFactory sqlSessionFactory =
    new SqlSessionFactoryBuilder().build(inputStream);
    //3.获取一个连接
    SqlSession session = sqlSessionFactory.openSession();
    //4.获取执行 SQL 的代理对象
    CustomerMapper customerMapper= session.getMapper(CustomerMapper.class);
    //5.执行 SQL
    Customer customer = new Customer();
    customer.setId(2);
    customer.setUsername("赵六");
    customer.setCity("天津");

    SimpleDateFormat sdf = new SimpleDateFormat("yyyy-MM-dd");
    Date date = sdf.parse("1998-10-02");
    customer.setBirthday(date);
    int result = customerMapper.modifyCustomer(customer);
    //6.处理结果
    if(result > 0){
        System.out.println("修改成功");
    }else{
        System.out.println("修改失败");
    }
    //7.提交事务
    session.commit();
    //8.关闭资源
    session.close();
}
```

测试执行结果如图 2-8 所示，客户信息修改成功。

```
Created connection 403147759.
Setting autocommit to false on JDBC Connection [com.mysql.jdbc.JDBC4Connection@18078bef]
==>  Preparing: update t_customer set username = ?,city = ?,birthday = ? where id = ?
==> Parameters: 赵六(String), 天津(String), 1998-10-02 00:00:00.0(Timestamp), 2(Integer)
<==    Updates: 1
修改成功
Committing JDBC Connection [com.mysql.jdbc.JDBC4Connection@18078bef]
Resetting autocommit to true on JDBC Connection [com.mysql.jdbc.JDBC4Connection@18078bef]
Closing JDBC Connection [com.mysql.jdbc.JDBC4Connection@18078bef]
Returned connection 403147759 to pool.
```

图 2-8  修改客户信息的结果

## 本章小结

在本章中，我们对 MyBatis 进行了初步的探索。作为 Java 生态系统中非常受欢迎的持久层框架，MyBatis 的灵活性和与 SQL 之间的紧密结合给开发者带来了很大的方便。本章主要介绍了以下内容。

(1) MyBatis 的核心概念与特性。

首先介绍了 MyBatis 的基本定义、功能以及其在 Java 中的定位，然后通过对比传统的 JDBC 和其他 ORM 框架，明确了 MyBatis 的优势所在。

(2) 环境搭建与基本配置。

通过简单的步骤，学会了如何搭建 MyBatis 的开发环境。对 MyBatis 的核心配置文件及其结构有了基本了解。

(3) 基础的 CRUD 操作。

了解了如何使用 MyBatis 进行数据的基本操作，如插入数据、查询数据、修改数据和删除数据。

## 课后习题

一、选择题

1. MyBatis 是一种(　　)的持久层框架。
   A. ORM 框架　　　B. Spring 框架　　　C. JavaScript 库　　　D. 数据库管理系统
2. 要在 Java 项目中使用 MyBatis，需要配置(　　)文件来定义 SQL 映射。
   A. .xml　　　　　B. .properties　　　C. .json　　　　　　D. .java
3. MyBatis 的核心目标之一是(　　)。
   A. 自动化数据库管理　　　　　　　B. 简化 Java 应用程序的部署
   C. 简化 SQL 编写　　　　　　　　D. 提供图形用户界面

二、填空题

1. 要搭建 MyBatis 开发环境，需要包括 MyBatis 的依赖项在项目的_____文件中。
2. 在 MyBatis 中，SQL 语句通常是在_____文件中定义的。

三、判断题

1. MyBatis 是一个基于 Hibernate 的 Java 持久化框架。　　　　　　　　　　　　(　　)
2. MyBatis 提供了自动化的数据库管理功能，无须编写 SQL 语句。　　　　　　　(　　)

四、简答题

1. 请叙述 MyBatis 的工作原理，并描述 MyBatis 是如何将 Java 对象映射到数据库表。
2. 在 MyBatis 中，SQL 映射文件的作用是什么？

五、实操题

创建一个 MyBatis 项目,实现与一个数据库表的 CRUD 操作。

数据的伦理与责任

数据是现代社会运转的"新石油"。MyBatis 使我们能更便捷地操作数据库,但同时也带来了一系列伦理和法律责任。我们需要遵循数据保护的法律规定,确保个人隐私和数据安全,防止数据滥用和泄露。

# 第 3 章
# 深入使用 MyBatis 框架

学习目标

1. 熟悉 MyBatis 的动态 SQL 功能：理解并能够熟练使用各种动态 SQL 标签来适应各种查询需求。

2. 掌握 MyBatis 的高级映射和查询方法：能够处理复杂的数据库关系和业务场景。

学习要点

1. 学习动态 SQL 深入：动态条件查询、批量操作等的实现。深入理解 <choose>、<when>、<foreach> 等高级标签的用法。

2. 掌握高级映射与查询：复杂关系的映射，如多对多、一对多等。

本章知识点结构如图 3-1 所示。

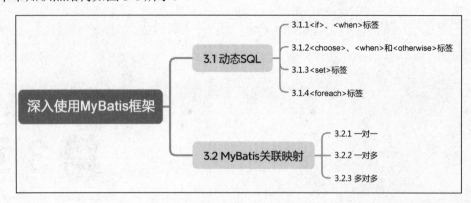

图 3-1 深入使用 MyBatis 框架

第 2 章介绍了 MyBatis 的基本使用，完成了对客户表的增删改查等操作，本章继续学习 MyBatis 的高级特性，即动态 SQL 和多表关联映射。

## 3.1 动态 SQL

在 MyBatis 中，动态 SQL 具有能够根据不同的条件生成不同 SQL 语句的功能。这样可以构建更加灵活且根据实际参数来动态更改的查询，而不是为每一种可能的查询条件编写一个固定的 SQL 语句。动态 SQL 在处理具有可选过滤条件的查询时尤为有用。

动画：动态 SQL 的重要性及实际应用

（1）在 IDEA 中创建一个新的项目 chapter03，将第 2 章 pom.xml 的依赖、实体类、mybatis-config.xml、Mapper 文件复制到当前项目中。

（2）在第 2 章中，实现数据的增删改查测试时，每一次都写一部分重复的代码创建 MyBatis 的 SqlSession 对象：

微课：MyBatisUtils 工具类开发

```
//1.加载配置文件信息
String resource = "mybatis-config.xml";
InputStream inputStream = Resources.getResourceAsStream(resource);
//2.创建一个连接池
SqlSessionFactory sqlSessionFactory =
    new SqlSessionFactoryBuilder().build(inputStream);
//3.获取一个连接
SqlSession session = sqlSessionFactory.openSession();
```

在实际的开发过程中，当一段代码在多个地方重复出现的时候，一个好的经验就是把这一段代码定义为一个函数以供重复调用，而不是任由相同的代码段在程序中多次出现。所以本节首先将上述重复的代码重构为一个函数，这样后续的开发直接调用此函数，而不必再编写重复代码。

在 com.demo 目录下新建一个 util 包，在该包下创建 MyBatisUtils 工具类，代码如下：

```
import org.apache.ibatis.io.Resources;
import org.apache.ibatis.session.SqlSession;
import org.apache.ibatis.session.SqlSessionFactory;
```

```java
import org.apache.ibatis.session.SqlSessionFactoryBuilder;
import java.io.InputStream;
public class MyBatisUtils {
    private static SqlSessionFactory sqlSessionFactory;
    static{
        try {
            //1.加载配置文件信息
            String resource = "mybatis-config.xml";
            InputStream inputStream = Resources.getResourceAsStream(resource);
            //2.创建一个连接池
            sqlSessionFactory =
                new SqlSessionFactoryBuilder().build(inputStream);
        }catch (Exception e){
            e.printStackTrace();
        }
    }
    public static SqlSession getSqlSession(){
        return sqlSessionFactory.openSession();
    }
}
```

以上代码在 MyBatisUtils 工具类中定义了一个方法 getSqlSession()，用以读取 MyBatis 配置文件，并创建 SqlSession 对象。在本章后续内容中，都将使用此方法获取 SqlSession 对象。

## 3.1.1 \<if\>、\<where\>标签

微课：if 和 where 标签

\<if\>标签允许进行条件判断，只有当条件满足时，包含在其内部的 SQL 片段才会被插入最终的 SQL 语句中。

\<where\>标签自动处理 WHERE 子句的 AND 和 OR 前缀，以确保在拼接 SQL 语句时得到的 SQL 语句是有效的，不会因为动态 SQL 的条件不满足而留下多余的 AND 或 OR。

在下面的示例中，程序通过客户姓名 username 和所在城市 city 查询客户信息列表，但是 username 和 city 并非都会在程序调用时提供值。代码中通过使用\<if\>和\<where\>标签来完成如果 username 参数为 null 而 city 参数不为 null 的查询。

(1) 在 CustomerMapper 接口中声明 queryByCondition()方法，参数是客户对象，用于存储要查询的条件数据。代码如下：

```java
public List<Customer> queryByCondition(Customer customer);
```

(2) 在 CustomerMapper.xml 文件中增加\<select\>标签，由于查询条件 username 和 city 可能存在，也可能不存在，所以使用\<if\>、\<where\>标签处理。代码如下：

```xml
<select id="queryByCondition" resultType="com.demo.pojo.Customer">
    select * from t_customer
    <where>
```

```
        <if test="username != null and username != '' ">
            username like concat('%',#{username},'%')
        </if>
        <if test="city != null and city != '' ">
            and city = #{city}
        </if>
    </where>
</select>
```

如果 username 参数不为 null，那么 SQL 中的 username = #{username}部分就会被包含在最终执行的 SQL 语句中。

(3) 在 TestMyBatis 测试类中添加 testQueryByCondition()方法，给 username 属性赋值，或者给 city 属性赋值，也可以都赋值或者都不赋值，分别执行后查看控制台中的 SQL 语句以及查询结果。代码如下：

```
@Test
public void testQueryByCondition() throws Exception {
    //1.获取一个连接
    SqlSession session = MyBatisUtils.getSqlSession();
    //2.获取执行 SQL 的代理对象
    CustomerMapper customerMapper= session.getMapper(CustomerMapper.class);
    //3.执行 SQL
    Customer ct = new Customer();
    //ct.setUsername("张");
    ct.setCity("北京");
    List<Customer> lists = customerMapper.queryByCondition(ct);
    //4.处理结果
    for (Customer customer : lists) {
        System.out.println(customer);
    }
    //5.关闭资源
    session.close();
}
```

(4) 例如当只给城市属性赋值为"北京"时，查询结果如图 3-2 所示。

```
==>  Preparing: select * from t_customer WHERE city = ?
==> Parameters: 北京(String)
<==    Columns: id, username, city, birthday
<==        Row: 1, 张三, 北京, 2001-10-08
<==        Row: 2, 李四, 北京, 2002-02-01
<==      Total: 2
Customer{id=1, username='张三', city='北京', birthday='Mon Oct 08 00:00:00 CST 2001'}
Customer{id=2, username='李四', city='北京', birthday='Fri Feb 01 00:00:00 CST 2002'}
```

图 3-2 查询结果

## 3.1.2 \<choose\>、\<when\>和\<otherwise\>标签

微课：choose、when、otherwise 标签

\<choose\>、\<when\>和\<otherwise\>标签是组合标签，类似于 Java 中的 switch-case-default 语句。\<choose\>标签与编程中的 switch 语句类似，它

为一组条件选择提供一个容器。<when>标签与编程中的 case 语句类似，它定义一个具体的条件，在<choose>标签内部使用，用于定义满足某个条件时的 SQL 片段。<otherwise>标签与编程中的 default 语句类似，它定义了当所有<when>条件都不满足时的默认行为，在<choose>标签内部，紧跟所有<when>标签之后使用。

下面这个需求是根据用户输入的条件查询客户信息。如果输入了姓名，则将姓名作为查询条件，否则将城市作为查询条件；如果姓名和城市都为空，则查询生日不为空的数据。

(1) 在 CustomerMapper 接口中定义一个 queryCustomerByNameAndCity()方法，参数 Customer 存储查询条件，返回值 List 集合存储查询结果。代码如下：

```java
public List<Customer> queryCustomerByNameAndCity(Customer customer);
```

(2) 在 CustomerMapper.xml 文件中增加对应的查询语句，查询条件使用动态 SQL 拼接完成。代码如下：

```xml
<select id="queryCustomerByNameAndCity" resultType="com.demo.pojo.Customer">
    select * from t_customer
    <where>
        <choose>
            <when test=" username != null and username != '' ">
                username like concat('%',#{username},'%')
            </when>
            <when test=" city != null and city != '' ">
                city = #{city}
            </when>
            <otherwise>
                birthday is not null
            </otherwise>
        </choose>
    </where>
</select>
```

(3) 在 TestMyBatis 测试类中增加 testQueryCustomerByNameAndCity()方法，创建一个客户对象，只给城市属性赋值为"上海"，则查询城市为上海的客户信息。代码如下：

```java
@Test
public void testQueryCustomerByNameAndCity() throws Exception {
    //1.获取一个连接
    SqlSession session = MyBatisUtils.getSqlSession();
    //2.获取执行SQL的代理对象
    CustomerMapper customerMapper= session.getMapper(CustomerMapper.class);
    //3.执行SQL
    Customer c1 = new Customer();
    c1.setCity("上海");
    List<Customer> lists = customerMapper.queryCustomerByNameAndCity(c1);
    //4.处理结果
    for (Customer customer : lists) {
        System.out.println(customer);
    }
    //5.关闭资源
    session.close();
}
```

(4) 由于 c1 对象没有给 username 属性赋值,所以 username 属性为空。只给 city 属性赋值,city 属性不为空,则根据城市名称进行查询,结果如图 3-3 所示。

```
==>  Preparing: select * from t_customer WHERE city = ?
==> Parameters: 上海(String)
<==    Columns: id, username, city, birthday
<==        Row: 3, 张丽, 上海, 2001-07-12
<==      Total: 1
Customer{id=3, username='张丽', city='上海', birthday='Thu Jul 12 00:00:00 CST 2001'}
```

图 3-3 查询结果

### 3.1.3 \<set\>标签

微课:set 标签

MyBatis 中,在处理更新(UPDATE)操作时,当不确定要更新哪些字段时,经常需要动态地设置字段的更新值,\<set\>标签正是为此而设计的。

第 2 章开发了修改客户信息的功能,如果在修改客户信息时客户的姓名不需要修改,也没有给客户姓名赋值,则会导致修改客户姓名为空的情况,这个问题可以使用 MyBatis 动态 SQL 的\<set\>和\<if\>标签解决。

(1) 将 2.3.4 小节中实现修改客户信息的 id 为 modifyCustomer 的 update 标签内容进行修改,使用\<set\>和\<if\>标签进行判断,当 username、city、birthday 不为空时再进行修改。代码如下:

```xml
<update id="modifyCustomer">
   update t_customer
   <set>
      <if test="username != null and username != '' ">
         username = #{username},
      </if>
      <if test="city != null and city != ''">
         city = #{city},
      </if>
      <if test="birthday != null">
         birthday = #{birthday}
      </if>
   </set>
   where id = #{id}
</update>
```

(2) 在测试类中只给 id 和 city 属性赋值。代码如下:

```java
@Test
public void testModify() throws Exception {
   //1.获取一个连接
   SqlSession session = MyBatisUtils.getSqlSession();
   //2.获取执行 SQL 的代理对象
   CustomerMapper customerMapper = session.getMapper(CustomerMapper.class);
   //3.执行 SQL
   Customer customer = new Customer();
```

```
customer.setId(3);
customer.setCity("武汉");

int result = customerMapper.modifyCustomer(customer);
//4.处理结果
if(result > 0){
    System.out.println("修改成功");
}else{
    System.out.println("修改失败");
}
//5.提交事务
session.commit();
//6.关闭资源
session.close();
}
```

在上面的示例中:
◎ 如果只传入了 username 参数,那么生成的 SQL 语句将为 UPDATE t_customer SET username = #{username} WHERE id = #{id}。
◎ 如果 username 和 city 都传入了参数,则生成的 SQL 语句将为 UPDATE t_customer SET username = #{username}, city= #{city} WHERE id = #{id}。

注意: 在每个 <if> 标签内部,我们都添加了一个逗号来分隔不同的更新字段。而 <set> 标签会自动处理这些逗号,确保不会出现多余的逗号。

(3) 由于我们在测试类的方法中只传入了 city 的名字,所以修改时只对 city 的名字进行修改,结果如图 3-4 所示。

```
==>  Preparing: update t_customer SET city = ? where id = ?
==>  Parameters: 武汉(String), 3(Integer)
<==     Updates: 1
修改成功
```

图 3-4　修改结果

## 3.1.4 <foreach>标签

在 MyBatis 中,<foreach>标签(见表 3-1)是一个强大的工具,用于处理集合,并能够将集合元素逐个插入到 SQL 语句中。这在处理 IN 条件或者批量插入、删除等场景中特别有用。

微课:foreach 标签

表 3-1　<foreach>标签的属性配置

| 属 性 名 | 属 性 值 |
| --- | --- |
| collection | 要迭代的集合的名称 |
| item | 每次迭代时,当前元素的名字 |
| index | 当前迭代的索引或键(如果是 Map 结构的话) |
| open | 整个集合的前缀 |
| close | 整个集合的后缀 |
| separator | 集合中每个元素之间的分隔符 |

(1) 在 CustomerMapper 接口中声明 removeCustomerByIds()方法，参数是 int 类型的数组，用于存储要删除客户的编号：

```java
public int removeCustomerByIds(@Param("ids") int[] ids);
```

(2) 在 CustomerMapper.xml 中增加<delete>标签，编写批量删除的 SQL 语句，<foreach>标签用于遍历集合或数组，collection="ids"指定了要遍历的集合。通常，这会对应到 Mapper 方法的参数。

- item="id"：当遍历集合时，每个元素都被称为 id。这是在<foreach>循环中用来引用当前元素的变量名。
- open="("和 close=")"：它们指定在生成片段的开始和结束处分别附加字符串。在这种情况下，它将生成的列表用括号括起来，这在 SQL 的 IN 子句中很常见。
- separator=","：这是生成的列表中每个项目之间使用的分隔符。对于 SQL 的 IN 子句，将用逗号分隔每个元素。

代码如下：

```xml
<delete id="removeCustomerByIds">
   delete from t_customer where id in
   <foreach collection="ids" item="id" open="(" close=")" separator=",">
      #{id}
   </foreach>
</delete>
```

(3) 在 TestMyBatis 类中定义一个批量删除的测试方法 testRemoveMany()，在测试类中创建一个数组，存储要批量删除客户的编号。代码如下：

```java
@Test
public void testRemoveMany() throws Exception{
    //1.获取一个连接
    SqlSession session = MyBatisUtils.getSqlSession();
    //2.获取执行SQL语句的代理对象
    CustomerMapper customerMapper= session.getMapper(CustomerMapper.class);
    //3.执行SQL语句
    int result = customerMapper.removeCustomerByIds(new int[]{1,2,3});
    //4.处理结果
    if(result > 0){
        System.out.println("删除成功");
    }else{
        System.out.println("删除失败");
    }
    //5.提交事务
    session.commit();
    //6.关闭资源
    session.close();
}
```

(4) 执行测试类中批量删除的方法，可以看到 where 条件中会拼接 3 个值，最后影响的行数为 3，代表实际删除了 3 条数据，如果表中没有对应的 id 值，实际影响的行数可能与图 3-5 所示的结果不一致，这未必代表错误，所以要以实际删除条数为准。

```
==> Preparing: delete from t_customer where id in ( ? , ? , ? )
==> Parameters: 1(Integer), 2(Integer), 3(Integer)
<==      Updates: 3
删除成功
```

图 3-5　删除结果

## 3.2　MyBatis 关联映射

动画：MyBatis
关联映射之路

在关系型数据库设计中，我们通常需要在不同的表之间建立关系，以保持数据的一致性和完整性。这些关系反映了现实世界中各个实体及其交互方式，通过理解这些关系，我们能够创建出更加有效的数据库架构，以满足应用程序的数据需求。

在复杂的业务场景下，数据通常会被分解并存储在多个表中，这些表通过一对一、一对多和多对多这三种常见的关系进行连接。这些关系不仅映射了现实世界的复杂性，也为数据提供了丰富的结构和语义。

(1) 一对一关系：这种关系描述的是两个表中的记录一一对应。例如，我们有一个用户表(User)和身份证信息表(IDCard)。每个用户在系统中都有一个唯一的记录，同时每个用户都对应一张唯一的身份证，反过来，每张身份证也只能对应一个用户。因此，用户表和身份证表之间就形成了一对一的关系。

(2) 一对多关系：这种关系描述的是一个表中的单个记录对应另一个表中的多个记录。例如，我们有一个作者表(Author)和书籍表(Book)。一位作者可以写多本书，因此，作者表中的一条记录(一个作者)可以对应书籍表中的多条记录(多本书)，这就形成了一对多的关系。但是，每本书只能有一个作者，所以反过来就不成立，这也是一对多关系的特点。

(3) 多对多关系：这种关系描述的是一个表中的记录可以对应另一个表中的多条记录，反过来，另一个表中的记录也可以对应第一个表中的多条记录。例如，我们有一个学生表(Student)和课程表(Course)。一个学生可以选多门课程，同时一门课程也可以被多个学生选学，所以学生表和课程表之间就形成了多对多的关系。多对多的关系通常需要通过一个中间表来进行管理，例如，我们可以通过一个学生课程表(StudentCourse)来记录哪个学生选了哪门课程，从而实现多对多的关系管理。

理解了表与表之间的关系之后，我们就可以利用这些关系来设计数据库，并编写更有效的 SQL 查询。通过 MyBatis，我们可以灵活地实现这些关联映射，从而简化数据库操作，提高应用程序的性能和可维护性。

### 3.2.1　一对一

一对一关系是指两个表中的每一行都唯一地对应另一张表中的一行。　微课：一对一查询

例如，以下 SQL 语句所示的用户表和身份证信息表之间，就是一对一的关系。每个用户(User)有编号 id 和姓名 username 属性，每个用户对应一个唯一身份证信息表(IdCard)。身份证信息表包含唯一编号 id、身份证号 number 以及对应的用户编号 id。

```
CREATE TABLE User (
  id INT PRIMARY KEY,
```

```
  username VARCHAR(255)
);

CREATE TABLE IdCard (
  id INT PRIMARY KEY,
  number VARCHAR(255),
  user_id INT,
  FOREIGN KEY (user_id) REFERENCES User (id)
);
```

使用以上SQL语句创建好数据表后,使用如下SQL语句向数据表中插入测试数据:

```
INSERT INTO User (id, username) VALUES (1, 'User1');
INSERT INTO IdCard (id, number, user_id) VALUES (1, '123456789', 1);
```

在MyBatis中,我们可以使用<association>标签来实现一对一关系的映射。<association>标签常见的属性包括property、javaType、column、select、fetchType、resultMap等,具体介绍如表3-2所示。

表3-2 <association>标签属性

| 属性名 | 用途 |
| --- | --- |
| property | 必填,对应Java模型类中的属性名 |
| javaType | 可选,Java模型类的全限定名或别名 |
| column | 可选,对应数据库表中的列名 |
| select | 可选,用于指定加载关联对象的SQL语句的id |
| fetchType | 可选,加载类型,包括lazy(懒加载)和eager(立即加载) |
| resultMap | 可选,外部结果映射的id |

(1) 在com.demo.pojo包下新建一个IdCard实体类。代码如下:

```
public class IdCard {

    private Long id;
    private String number;

    //省略getter和setter、toString

}
```

(2) 在com.demo.pojo包下新建一个User实体类。代码如下:

```
public class User {
  private Long id;
  private String username;
  private IdCard idCard;

  //省略getter和setter、toString
}
```

(3) 在com.demo.mapper包下新建一个UserMapper接口。代码如下:

```java
public interface UserMapper {

    public List<User> queryUser();

}
```

(4) 在 resources 文件夹的 mappers 目录下新建一个 UserMapper.xml 文件。代码如下：

```xml
<?xml version="1.0" encoding="UTF-8"?>
<!DOCTYPE mapper
      PUBLIC "-//mybatis.org//DTD Mapper 3.0//EN"
      "http://mybatis.org/dtd/mybatis-3-mapper.dtd">
<mapper namespace="com.demo.mapper.UserMapper">

    <resultMap id="userResultMap" type="User">
        <id property="id" column="id" />
        <result property="username" column="username" />
        <association property="idCard" javaType="IdCard">
            <id property="id" column="id" />
            <result property="number" column="number" />
        </association>
    </resultMap>

    <select id="queryUser" resultMap="userResultMap">
        SELECT u.*, i.id, i.number
        FROM User u
        LEFT JOIN IdCard i ON u.id = i.user_id
    </select>

</mapper>
```

(5) 在 mybatis-config.xml 配置文件中增加<typeAliases>标签，配置实体类的别名，Mapper 文件中的 type 属性可以简写为 User，在<mappers>标签中注册 UserMapper.xml，代码如下：

```xml
<?xml version="1.0" encoding="UTF-8" ?>
<!DOCTYPE configuration PUBLIC "-//mybatis.org//DTD Config 3.0//EN"
      "http://mybatis.org/dtd/mybatis-3-config.dtd">
<configuration>

    <settings>
        ...
    </settings>

    <typeAliases>
        <typeAlias type="com.demo.pojo.User" alias="User"></typeAlias>
        <typeAlias type="com.demo.pojo.IdCard" alias="IdCard"></typeAlias>
    </typeAliases>

    <environments default="development">
        ...
    </environments>
```

```xml
    <mappers>
        <mapper resource="mappers/CustomerMapper.xml"/>
        <mapper resource="mappers/UserMapper.xml"/>
    </mappers>
</configuration>
```

(6) 在测试类中编写 testQueryUser()方法，对查询用户的方法进行测试。代码如下：

```java
@Test
public void testQueryUser() throws Exception{
    //1.获取一个连接
    SqlSession session = MyBatisUtils.getSqlSession();
    //2.获取执行SQL语句的代理对象
    UserMapper userMapper = session.getMapper(UserMapper.class);
    //3.执行SQL语句
    List<User> lists = userMapper.queryUser();
    //4.处理结果
    for (User user : lists) {
        System.out.println(user);
    }
    //5.关闭资源
    session.close();
}
```

(7) 执行测试代码，查看控制台的查询结果，如图 3-6 所示。

```
==>  Preparing: SELECT u.*, i.id, i.number FROM User u LEFT JOIN IdCard i ON u.id = i.user_id
==>  Parameters:
<==     Columns: id, username, id, number
<==         Row: 1, User1, 1, 123456789
<==       Total: 1
User{id=1, username='User1', idCard=IdCard{id=1, number='123456789'}}
```

图 3-6 一对一查询结果

## 3.2.2 一对多

一对多关系是指一个表中的一行对应另一张表中的多行。

例如，以下 SQL 语句的作者信息表和图书信息表之间，就是一对多的关系。每个作者(Author)有编号 id 和姓名 name 属性，每个作者可以出版多本图书，所以每个作者信息对应多个图书信息的信息表(Book)。图书信息表中包含唯一编号 id、书名 title 以及对应作者编号 id。建表语句如下：

```sql
CREATE TABLE Author (
  id INT PRIMARY KEY,
  name VARCHAR(255)
);

CREATE TABLE Book (
  id INT PRIMARY KEY,
```

```
    title VARCHAR(255),
    author_id INT,
    FOREIGN KEY (author_id) REFERENCES Author(id)
);
```

使用以上 SQL 语句创建好数据表后,使用如下 SQL 语句向数据表中插入测试数据:

```
INSERT INTO Author (id, name) VALUES (1, 'Author1');
INSERT INTO Book (id, title, author_id) VALUES (1, 'Book1', 1);
INSERT INTO Book (id, title, author_id) VALUES (2, 'Book2', 1);
```

在 MyBatis 中,我们可以使用<collection>标签来实现一对多关系的映射。<collection>标签常见的属性和<association>标签类似,但多了一个 ofType 属性,用于指定集合元素的类型。

(1) 在 com.demo.pojo 目录下创建 Book、Author 实体类。代码如下:

```java
public class Book {

    private Long id;
    private String title;

    //省略 getter、setter 和 toString

}

public class Author {

    private Long id;
    private String name;
    private List<Book> books;

    //省略 getter、setter 和 toString

}
```

(2) 创建 AuthorMapper 接口文件。代码如下:

```java
public interface AuthorMapper {

    public List<Author> queryAuthor();

}
```

(3) 在 resources 文件夹的 mappers 目录下创建 AuthorMapper.xml 文件。代码如下:

```xml
<?xml version="1.0" encoding="UTF-8"?>
<!DOCTYPE mapper
        PUBLIC "-//mybatis.org//DTD Mapper 3.0//EN"
        "http://mybatis.org/dtd/mybatis-3-mapper.dtd">

<mapper namespace="com.demo.mapper.AuthorMapper">
```

```xml
<resultMap id="authorResultMap" type="Author">
    <id property="id" column="id" />
    <result property="name" column="name" />
    <collection property="books" ofType="Book">
        <id property="id" column="book_id" />
        <result property="title" column="title" />
    </collection>
</resultMap>

<select id="queryAuthor" resultMap="authorResultMap">
    SELECT a.*, b.id as book_id, b.title
    FROM Author a
    LEFT JOIN Book b ON a.id = b.author_id
</select>
</mapper>
```

注意：记得在 mybatis-config.xml 文件中增加别名、注册 Mapper 文件的配置。

(4) 在测试类中编写 testQueryAuthorAndBooks()方法，对查询作者的方法进行测试。代码如下：

```java
@Test
public void testQueryAuthorAndBooks() throws Exception {
    //1.获取一个连接
    SqlSession session = MyBatisUtils.getSqlSession();
    //2.获取执行SQL的代理对象
    AuthorMapper authorMapper = session.getMapper(AuthorMapper.class);
    //3.执行SQL
    List<Author> lists = authorMapper.queryAuthor();
    //4.处理结果
    for (Author author : lists) {
        System.out.println(author);
    }
    //5.关闭资源
    session.close();
}
```

(5) 运行程序，结果如图 3-7 所示。

```
==>  Preparing: SELECT a.*, b.id as book_id, b.title FROM Author a LEFT JOIN Book b ON a.id = b.author_id
==> Parameters:
<==    Columns: id, name, book_id, title
<==        Row: 1, Author1, 1, Book1
<==        Row: 1, Author1, 2, Book2
<==      Total: 2
Author{id=1, name='Author1', books=[Book{id=1, title='Book1'}, Book{id=2, title='Book2'}]}
```

图 3-7　一对多查询结果

## 3.2.3　多对多

微课：多对多查询

多对多关系是指一个表中的一行对应另一张表中的多行，反过来也是如此。多对多关系通常通过第三个表(连接表)来实现。

例如，以下 SQL 语句的学生和课程之间就是多对多的关系。学生信息表 Student 包含学生编号 id 和姓名 name。课程信息表 Course 包含编号 id 和名称 title。学生和课程的多对多关系由 Student_Course 表维护。Student_Course 表中的 student_id 和 course_id 分别对应 Student 表的 id 和 Course 表的 id。建表语句如下。

```sql
CREATE TABLE Student (
  id INT PRIMARY KEY,
  name VARCHAR(255)
);

CREATE TABLE Course (
  id INT PRIMARY KEY,
  title VARCHAR(255)
);

CREATE TABLE Student_Course (
  student_id INT,
  course_id INT,
  PRIMARY KEY (student_id, course_id),
  FOREIGN KEY (student_id) REFERENCES Student(id),
  FOREIGN KEY (course_id) REFERENCES Course(id)
);
```

使用以上 SQL 语句创建好数据表后，使用如下 SQL 语句向数据表中插入测试数据：

```sql
INSERT INTO Student (id, name) VALUES (1, 'Student1');
INSERT INTO Student (id, name) VALUES (2, 'Student2');
INSERT INTO Course (id, title) VALUES (1, 'Course1');
INSERT INTO Course (id, title) VALUES (2, 'Course2');
INSERT INTO Student_Course (student_id, course_id) VALUES (1, 1);
INSERT INTO Student_Course (student_id, course_id) VALUES (1, 2);
INSERT INTO Student_Course (student_id, course_id) VALUES (2, 2);
```

在 MyBatis 中，通过组合使用<association>和<collection>标签来实现多对多关系的映射。
(1) 在 com.demo.pojo 目录下创建 Course、Student 实体类。代码如下：

```java
public class Course {

    private Long id;
    private String title;

    // 省略 getter、setter 和 toString

}

public class Student {

    private Long id;
    private String name;
    private List<Course> courses;
```

```
    // 省略 getter、setter 和 toString
}
```

(2) 创建 StudentMapper 接口文件。代码如下：

```
public interface StudentMapper {

    public List<Student> queryStudent();

}
```

(3) 在 resources 文件夹的 mappers 目录下创建 StudentMapper.xml 文件。代码如下：

```
<?xml version="1.0" encoding="UTF-8"?>
<!DOCTYPE mapper
        PUBLIC "-//mybatis.org//DTD Mapper 3.0//EN"
        "http://mybatis.org/dtd/mybatis-3-mapper.dtd">

<mapper namespace="com.demo.mapper.StudentMapper">
    <resultMap id="studentResultMap" type="Student">
        <id property="id" column="id" />
        <result property="name" column="name" />
        <collection property="courses" ofType="Course">
            <id property="id" column="course_id" />
            <result property="title" column="title" />
        </collection>
    </resultMap>

    <select id="queryStudent" resultMap="studentResultMap">
        SELECT s.*, c.id as course_id, c.title
        FROM Student s
        LEFT JOIN Student_Course sc ON s.id = sc.student_id
        LEFT JOIN Course c ON sc.course_id = c.id
    </select>
</mapper>
```

(4) 在测试类中编写 testQueryStudentAndcourses()方法，对查询学生的方法进行测试。代码如下：

```
@Test
public void testQueryStudentAndCourses() throws Exception {
    //1.获取一个连接
    SqlSession session = MyBatisUtils.getSqlSession();
    //2.获取执行 SQL 的代理对象
    StudentMapper studentMapper = session.getMapper(StudentMapper.class);
    //3.执行 SQL
    List<Student> lists = studentMapper.queryStudent();
    //4.处理结果
    for (Student student : lists) {
        System.out.println(student);
    }
```

```
    //5.关闭资源
    session.close();
}
```

(5) 执行测试代码，查看控制台的查询结果，如图3-8所示。

```
==>  Preparing: SELECT s.*, c.id as course_id, c.title FROM Student s LEFT JOIN Student_Course sc ON s.id = sc.student_id LEFT JOIN Course c ON sc.course_id
==>  Parameters:
<==  Columns: id, name, course_id, title
<==      Row: 1, Student1, 1, Course1
<==      Row: 1, Student1, 2, Course2
<==      Row: 2, Student2, 2, Course2
<==    Total: 3
```

图 3-8  多对多查询结果

## 本 章 小 结

在本章中，我们深入探讨了 MyBatis 的两个核心特性：动态 SQL 和表关联映射。

动态 SQL 为我们提供了一种灵活的方式，使得 SQL 语句可以根据传入的参数值动态地生成和修改，从而使我们的查询更具适应性。通过学习和应用<if>、<choose>、<when>、<otherwise>、<set>和<foreach>等标签，我们可以轻松地处理各种各样的查询需求，无论是单一条件的筛选，还是复杂数组参数的处理。

表关联映射则侧重于数据库中不同表之间的关系。我们探讨了如何利用 MyBatis 来映射一对一、一对多和多对多这三种常见的数据库关系，并通过具体的例子和代码片段，理解了如何进行实际的映射配置。这使得我们在处理复杂的数据库结构时，能够保持代码的简洁性和可读性。

总之，通过本章的学习，我们对 MyBatis 有了更深入的了解，不仅掌握了其基本使用方法，还学会了如何在实际项目中高效、灵活地利用 MyBatis 进行数据操作。

## 课 后 习 题

一、选择题

1. 在 MyBatis 中，一对多的关系通常通过(　　)标签来表示。
   A. <one-to-one>　　　B. <many-to-one>　　　C. <one-to-many>　　　D. <many-to-many>
2. 当在 MyBatis 中执行一对一查询时，通常需要使用(　　)标签来定义关联关系。
   A. <association>　　　B. <collection>　　　C. <result>　　　D. <select>
3. MyBatis 中的动态 SQL 可以用来实现(　　)功能。
   A. 编写静态 SQL 语句　　　　　　B. 根据条件动态生成 SQL 语句
   C. 执行存储过程　　　　　　　　D. 连接多个数据库

二、填空题

1. 在 MyBatis 中，动态 SQL 通常使用_____标签来实现根据条件生成 SQL 语句。
2. 一对多查询中，通常需要使用_____标签来定义集合属性的映射。

## 三、判断题

1. 在MyBatis中，多对多的关系无法直接映射到数据库表，需要使用中间表来表示。
（    ）

2. 动态SQL允许根据不同的条件生成不同的SQL语句，以适应不同的查询需求。
（    ）

## 四、简答题

1. 请解释MyBatis中的一对多(One-to-Many)查询是什么，以及如何在映射文件中配置一对多关系。

2. 动态SQL在MyBatis中的作用是什么？举例说明在实际应用中如何使用动态SQL来构建灵活的查询语句。

## 五、实操题

1. 在MyBatis中，<choose>、<when>和<otherwise>标签的组合通常用于处理什么样的查询场景？请实例说明。

2. 请描述如何在MyBatis中进行关联映射，并给出相应的XML配置片段。<foreach>标签在MyBatis中有哪些常见的应用场景，请提供一个实例来展示其用法。

### 高效的代码，高效的社会

深入学习和应用MyBatis不仅能提高代码的效率和质量，也是在社会主义现代化建设中追求效率和高质量发展的体现。掌握这一技术，意味着我们有能力更好地服务社会，推动社会进步。

# 第 4 章
# Spring 框架使用指南

**学习目标**

1. 理解 IoC(控制反转)和 AOP(面向切面编程)的基本概念及其在 Spring 框架中的应用。
2. 学会使用 XML 和注解配置 Spring 的 IoC 容器。
3. 熟悉 Spring AOP 的核心概念,包括切面、连接点、切点等。

**学习要点**

1. IoC 容器的基本概念和作用。
2. 基于 XML 的配置方法。
3. 基于注解的配置方法。
4. Bean 的作用域。
5. AOP 的核心概念:切面、连接点、切点。
6. 如何使用注解来定义切面。

本章知识点结构如图 4-1 所示。

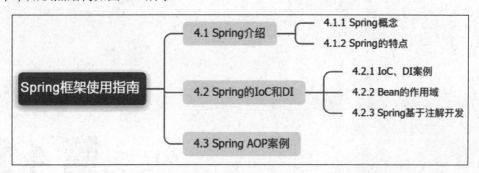

图 4-1　Spring 框架使用指南

## 4.1　Spring 介绍

### 4.1.1　Spring 概念

Spring 是一个开源框架，用于简化 Java 应用程序的开发、测试和部署过程，它为 JavaEE 应用提供了一个全面的编程和配置模型。自从 2003 年首次发布以来，Spring 已经成为 Java 开发者的首选框架，特别是企业级应用程序的开发。

动画：探索 Spring 框架——轻松管理企业级应用

### 4.1.2　Spring 的特点

**1. 轻量级**

Spring 是一个轻量级的框架，它的基本版本大约 2MB。

**2. 控制反转(IoC)**

Spring 通过 IoC 容器提供了控制反转，帮助开发者解耦应用程序组件。

**3. 面向切面编程(AOP)**

Spring 支持 AOP，允许定义横切关注点，并在应用程序的多个地方进行中心化管理。

**4. 声明式事务管理**

提供了对声明式事务管理的支持，使用户能够在不触及代码的情况下管理事务。

**5. 数据访问**

Spring 提供了与 JDBC 和 ORM 工具的集成。

**6. 模块化**

Spring 框架是模块化的，这意味着可以选择使用其中的哪些部分，而不必全部导入。这些模块为构建企业级应用程序提供了全面的支持。

Spring 是 JavaEE 开发的一站式解决方案，为开发者提供了一个简单而强大的方式来构建复杂的应用程序，Spring 图标如图 4-2 所示。随着时间的推移，Spring 社区持续地增加了许多子项目，如 SpringBoot、SpringData、SpringCloud 等，进一步增强了其在企业开发中的地位。

图 4-2　Spring 的图标

## 4.2　Spring 的 IoC 和 DI

### 4.2.1　IoC、DI 案例

控制反转(Inversion of Control，IoC) 是一个编程概念，其核心思想是将原本在程序中手动创建对象、管理对象之间的依赖关系等工作，交给外部容器完成。这意味着程序不再拥有流程的主控权，对象的生命周期和依赖关系由容器负责管理。

动画：Spring IoC 容器——让代码更优雅

在 Spring 框架中，IoC 容器就是这样的一个组件，它负责实例化、配置和组装 Bean。容器通过读取提供给它的配置元数据来获得指令(例如 XML、注解或 Java 配置)。

微课：Spring 的 IoC、DI

依赖注入(Dependency Injection，DI)是实现 IoC 的方法。它允许我们将依赖关系从应用程序的代码中移除，然后通过容器的方式来进行注入。

这样做有如下几个优点。

1. 解耦合

由于组件不再直接依赖于具体的实现，因此更容易替换或更新。

2. 可测试性

依赖可以通过接口或抽象类来定义，因此在测试时可以方便地使用模拟对象。

3. 可维护性

减少了代码之间的依赖关系，使得系统更为模块化，更易于维护和扩展。

下面通过实例来介绍 Spring 的 IoC 和 DI 的用法。具体步骤如下。

(1) 在 IntelliJ IDEA 中创建一个新的 Maven 项目 chapter04，添加 Spring 核心、JUnit 单元测试相关依赖，代码如下：

```xml
<dependencies>
    <!-- 导入 spring 的坐标 spring-context，对应版本是 5.2.10.RELEASE -->
    <dependency>
        <groupId>org.springframework</groupId>
        <artifactId>spring-context</artifactId>
        <version>5.2.10.RELEASE</version>
    </dependency>
```

```xml
<!-- 导入junit的测试包 -->
<dependency>
    <groupId>junit</groupId>
    <artifactId>junit</artifactId>
    <version>4.13.2</version>
    <scope>test</scope>
</dependency>
</dependencies>
```

(2) 在项目的 src/main/java 目录下创建 com.demo.dao 包，在该包下创建 BookDao 接口，声明添加图书方法 save()，代码如下：

```java
package com.demo.dao;

public interface BookDao {
    void save();
}
```

(3) 在项目的 src/main/java 目录下创建 com.demo.dao.impl 包，在该包下创建 BookDaoImpl 类，实现 BookDao 接口以及 save()方法，代码如下：

```java
package com.demo.dao.impl;

import com.demo.dao.BookDao;

public class BookDaoImpl implements BookDao {

    @Override
    public void save() {
        System.out.println("DAO 数据库访问层：BookDaoImpl 的 save 方法被调用");
    }

}
```

(4) 在项目的 src/main/java 目录下创建 com.demo.service 包，在该包下创建 BookService 接口，声明 save()方法，代码如下：

```java
package com.demo.service;

public interface BookService {

    void save();
}
```

(5) 在项目的 src/main/java 目录下创建 com.demo.service.impl 包，在该包下创建 BookServiceImpl 类，实现 BookService 接口以及 save()方法。在 BookServiceImpl 类中添加 BookDao 类型的属性，并且生成对应的 save()方法，代码如下：

```java
package com.demo.service.impl;

import com.demo.dao.BookDao;
```

```java
import com.demo.service.BookService;

public class BookServiceImpl implements BookService {

    private BookDao bookDao;

    public void save() {
        System.out.println("Service 业务逻辑层: BookServiceImpl 的 save 方法被调用");
        bookDao.save();
    }

    public void setBookDao(BookDao bookDao) {
        this.bookDao = bookDao;
    }
}
```

（6）在 resources 目录下创建 applicationContext.xml 文件，配置 service 和 dao 对应的 Bean 对象，并且在 service 中注入 dao 对象，代码如下：

```xml
<?xml version="1.0" encoding="UTF-8"?>
<beans xmlns="http://www.springframework.org/schema/beans"
    xmlns:xsi="http://www.w3.org/2001/XMLSchema-instance"
    xsi:schemaLocation="http://www.springframework.org/schema/beans
        http://www.springframework.org/schema/beans/spring-beans.xsd">


    <!--
    Bean 标签：表示配置 Bean
    id 属性：表示给 Bean 起名字
    class 属性：表示给 Bean 定义类型
    -->
    <bean id="bookDao" class="com.demo.dao.impl.BookDaoImpl"/>

    <bean id="bookService" class="com.demo.service.impl.BookServiceImpl">
        <!-- 配置 service 与 dao 的关系
            property 标签：表示配置当前 Bean 的属性
            name 属性：表示配置哪一个具体的属性
            ref 属性：表示参照哪一个 Bean
        -->
        <property name="bookDao" ref="bookDao"/>
    </bean>

</beans>
```

上面的配置中定义了两个 Spring Bean：bookDao 和 bookService，其中 bookService 中会引用 bookDao。当 Spring 程序启动时，Spring 会加载此配置文件，然后根据 Bean 的定义创建对应的 Java 对象。

本例中，Spring 会创建两个 Java 对象，使用类 com.demo.dao.impl.BookDaoImpl 创建名为 bookDao 的对象，使用类 com.demo.service.impl.BookServiceImpl 创建名为 bookService 的对象，并且在创建 bookService 的时候，会将其中的 bookDao 属性赋值给刚刚创建的

bookDao 对象。因此，在 bookService 中只需要声明 bookDao 字段，其创建与维护则交由 Spring。

（7）在项目的 src/test/java 目录下创建 com.demo.test 包。在该包下创建 TestBean 类，在该测试类的主函数中，读取 Spring 的配置文件，创建 ApplicationContext 对象，获取 BookService 的对象，然后调用对象中的方法，代码如下：

```java
package com.demo.test;

import com.demo.service.BookService;
import org.springframework.context.ApplicationContext;
import org.springframework.context.support.ClassPathXmlApplicationContext;

public class TestBean {
    public static void main(String[] args) {
        //1.创建IoC容器对象,加载Spring核心配置文件
        ApplicationContext ctx =
            new ClassPathXmlApplicationContext("applicationContext.xml");
        //2.从IoC容器中获取Bean对象(BookService对象)
        BookService bookService= (BookService)ctx.getBean("bookService");
        //3.调用Bean对象(BookService对象)的方法
        bookService.save();
    }

}
```

（8）运行测试程序，通过输出信息可以看到 bookService 和 bookDao 的对应方法被调用了，如图 4-3 所示。

Service业务逻辑层：BookServiceImpl的save方法被调用
DAO数据库访问层：BookDaoImpl的save方法被调用

图 4-3　Spring IoC 运行效果

### 4.2.2　Bean 的作用域

微课：Spring 的 Bean 作用域

<bean>标签的 scope 属性可以配置为 singleton 或者 prototype。

**1. singleton**

每个 Spring IoC 容器返回的是相同的 Bean 实例。无论请求多少次，都会得到同一个对象实例。

**2. prototype**

每次请求都会创建一个新的 bean 实例。

singleton 是默认的 Bean 创建方式。下面通过一个例子来演示其效果。

（1）配置一个 JavaBean 为 singleton，代码如下：

```xml
<bean id="bookService"
    class="com.demo.service.impl.BookServiceImpl"
```

```
      scope="singleton">
        <property name="bookDao" ref="bookDao"/>
</bean>
```

(2) 创建一个 TestScope 测试类，从 Spring 容器中获取 bookService 对象，并且分别打印对象地址，代码如下：

```
package com.demo.test;

import com.demo.service.BookService;
import org.springframework.context.ApplicationContext;
import org.springframework.context.support.ClassPathXmlApplicationContext;

public class TestScope {

    public static void main(String[] args) {
        //1.创建 IoC 容器对象，加载 Spring 核心配置文件
        ApplicationContext ctx =
        new ClassPathXmlApplicationContext("applicationContext.xml");
        //2.从 IoC 容器中获取 Bean 对象(BookService 对象)
        BookService bookService1
        = (BookService)ctx.getBean("bookService");
        BookService bookService2
         = (BookService)ctx.getBean("bookService");
        //3.打印 service 对象的内存地址
        System.out.println(" bookService1 = " + bookService1);
        System.out.println(" bookService2 = " + bookService2);
    }

}
```

(3) 对 scope 分别设置不同的属性值，测试结果如图 4-4、图 4-5 所示。

图 4-4　scope 设置为 singleton

图 4-5　scope 设置为 prototype

由以上结果可知，当 Bean 的 scope 属性是 singleton 时，在测试代码中获取了两次 BookService 类的对象，这两次获取的其实是同一个对象。当 Bean 的 scope 的属性是 prototype 时，在测试代码中获取了两次 BookService 类的对象，这两次获取时分别创建了不同的对象。

### 4.2.3 Spring 基于注解开发

在以上的内容中，都是通过 XML 方式来对 Spring 进行配置。Spring 还提供了另外一种配置方式，即基于注解的配置方式。相对于传统的 XML 方式，Spring 基于注解的开发带来了许多优势，它使得代码更简洁、更具有表达性，并且减少了大量的冗余配置。

微课：Spring 基于注解开发

下面就通过一个例子来介绍 Spring 基于注解的配置方式。

(1) 在项目的 src/main/java 目录下创建 com.demo.config 包，在该包下创建 SpringConfig 类，代码如下：

```java
import org.springframework.context.annotation.ComponentScan;
import org.springframework.context.annotation.Configuration;

@Configuration
@ComponentScan(value = "com.demo")
public class SpringConfig {
}
```

@Configuration 是 Spring 提供的核心注解，主要用于标识一个类来充当 Spring 的配置，并定义 Bean 的配置信息。这个注解使我们可以通过编写 Java 代码而非传统的 XML 来配置 Spring。

@ComponentScan 注解用来指定 Spring 扫描哪些包以查找标记为组件的类(如 @Component、@Service、@Repository、@Controller 等)。

(2) 在 BookDaoImpl 类、BookServiceImpl 类上添加@Component 注解，代码如下：

```java
@Component("bookDao")
public class BookDaoImpl implements BookDao {

}

@Component("bookService")
public class BookServiceImpl implements BookService {

    @Autowired
    private BookDao bookDao;

}
```

@Component 是 Spring 提供的基本注解，用于定义一个 Bean。其目的是让 Spring 扫描类时能够识别并注册为一个 Spring Bean。当标记一个类为 @Component 时，Spring 会自动检测这个类，并将其添加到 Spring 应用上下文中。

@Component("bookDao") 中的 bookDao 参数是可选的，指定了这个 Bean 的 ID。如果没有指定，Spring 会默认使用首字母小写的类名作为 Bean 的 ID。

@Component 有几个衍生的注解，这些注解在特定的场景中有特定的含义，但它们的

基本行为和 @Component 是一样的。这些派生注解包括@Service、@Repository 和 @Controller。通常情况下，它们被用于层次化地标记组件，提供更多的上下文信息。@Service 标记在服务层组件上，@Repository 标记在数据访问组件上，即 DAO 组件。@Controller 标记在表示层组件上，即 MVC 控制器上。

(3) 创建一个测试类，对基于注解开发进行测试，代码如下：

```
package com.demo.test;

import com.demo.config.SpringConfig;
import com.demo.service.BookService;
import org.springframework.context.ApplicationContext;
import org.springframework.context.annotation.
nnotationConfigApplicationContext;

public class TestAnnotation {
    public static void main(String[] args) {
        //1.创建 IoC 容器对象，加载 Spring 配置类
        ApplicationContext ctx =
            new AnnotationConfigApplicationContext(SpringConfig.class);
        //2.从 IoC 容器中获取 Bean 对象(BookService 对象)
        BookService bookService= (BookService)ctx.getBean("bookService");
        //3.调用 Bean 对象(BookService 对象)的方法
        bookService.save();
    }
}
```

运行测试类，结果如图 4-6 所示。

图 4-6　注解开发测试结果

## 4.3　Spring AOP 案例

### 1. AOP 简介

面向切面编程(AOP)是一种编程范式，它提供了将切面(关注点)与它们所影响的程序的其余部分(通常称为业务逻辑或核心关注点)分离的能力。在这种方式下，切面可以定义为跨多个方法或对象的重复代码。AOP 主要是在企业级应用中进行切面的模块化，这些切面有时跨越多个模块。

AOP 的主要优势有以下几个。
1) 代码分离
AOP 帮助用户将业务逻辑与系统服务(如日志记录、事务管理等)分离。
2) 提高模块性
可以将多个应用程序模块中的关注点抽象为一个切面。
3) 提高维护性
切面中心化的行为意味着更少的代码重复，从而更易于维护。

动画：探索 Spring AOP——编程的艺术

### 2. AOP 案例

本节将通过一个实例来介绍 Spring AOP 的使用方法。在一些场景中，需要在某些方法执行之前和之后执行一些操作，例如统计业务逻辑层方法的执行时间。通常的方式是在执行方法前记录当前时间，然后在执行方法后再次记录当前时间，最后用后面的时间减去前面的时间，就可以实现此功能。但是如果需要统计的地方很多的话，在程序中将出现很多次类似的重复代码，并且这些统计代码实际上与业务逻辑并不相关。在这种情况下，使用 Spring AOP 来实现此类功能是很合适的。其详细步骤如下。

微课：AOP

(1) 在项目的 pom.xml 中添加 AOP 开发所需的依赖，代码如下：

```xml
<dependency>
    <groupId>org.aspectj</groupId>
    <artifactId>aspectjweaver</artifactId>
    <version>1.9.4</version>
</dependency>

<dependency>
    <groupId>org.springframework</groupId>
    <artifactId>spring-test</artifactId>
    <version>5.2.10.RELEASE</version>
    <scope>test</scope>
</dependency>
```

(2) 定义通知类，然后定义切入点，代码如下：

```java
package com.demo.advice;

import org.aspectj.lang.ProceedingJoinPoint;
import org.aspectj.lang.annotation.*;
import org.springframework.stereotype.Component;

@Component
@Aspect
public class MyAdvice {

    @Pointcut("execution(void com.demo.service.*.*(..))")
    private void pt(){}

    @Around("pt()")
    public void aroundMethod(ProceedingJoinPoint pjp) throws Throwable {
        System.out.println("环绕通知 执行目标方法之前");
```

```
        long start = System.currentTimeMillis();
        pjp.proceed();
        long end = System.currentTimeMillis();
        System.out.println("环绕通知 执行目标方法之后");
        System.out.println( "耗时" + (end-start) + "毫秒");
    }
}
```

(3) 设置 Spring 对 AOP 注解的支持，代码如下：

```java
package com.demo.config;

import org.springframework.context.annotation.ComponentScan;
import org.springframework.context.annotation.Configuration;
import org.springframework.context.annotation.EnableAspectJAutoProxy;

@Configuration
@ComponentScan(value = "com.demo")
@EnableAspectJAutoProxy
public class SpringConfig {
}
```

(4) 编写 AOP 案例测试类，代码如下：

```java
package com.demo.test;

import com.demo.config.SpringConfig;
import com.demo.dao.BookDao;
import com.demo.service.BookService;
import org.junit.Test;
import org.junit.runner.RunWith;
import org.springframework.beans.factory.annotation.Autowired;
import org.springframework.test.context.ContextConfiguration;
import org.springframework.test.context.junit4.SpringJUnit4ClassRunner;

@RunWith(SpringJUnit4ClassRunner.class)
@ContextConfiguration(classes = SpringConfig.class)
public class TestAOP {

    @Autowired
    private BookService bookService;

    @Test
    public void testAop() {
        bookService.save();
    }

}
```

(5) 运行测试类，结果如图 4-7 所示，在 bookService.save() 方法执行的前后打印了对应的日志，并且统计出了方法的执行时间。

图 4-7 AOP 测试结果

## 本 章 小 结

本章主要介绍了 Spring 的两个核心概念：IoC 和 AOP。IoC 解决了对象之间依赖关系的管理问题，使得代码更加灵活和可维护。AOP 则解决了如何有效地处理分布在多个对象中的横切关注点问题。

## 课 后 习 题

一、选择题

1. 在 Spring 框架中，IoC 的全称是(　　)。
   A. Inversion of Control　　　　　　B. Input of Control
   C. Inheritance of Classes　　　　　　D. Interface of Classes
2. IoC 容器是 Spring 中的一个重要概念，它的主要作用是(　　)。
   A. 管理数据库连接池　　　　　　B. 管理应用程序的业务逻辑
   C. 管理对象的生命周期和依赖关系　　D. 管理用户会话信息
3. AOP(面向切面编程)是用来(　　)的。
   A. 管理数据库操作　　　　　　B. 处理日志和事务
   C. 创建用户界面　　　　　　　D. 进行数据分析

二、填空题

1. 在 Spring 框架中，IoC 容器通常由一个或多个 XML 文件来配置，其中最重要的配置标签是_____。
2. AOP 允许将通用的横切关注点(cross-cutting concerns)从应用程序的核心逻辑中分离出来，通常通过使用_____进行定义和管理。

三、判断题

1. 在 Spring 框架中，IoC 容器负责创建和管理对象的依赖关系。　　　　　　(　　)
2. AOP 是一种用于创建用户界面的技术。　　　　　　　　　　　　　　　(　　)

### 四、简答题

1. 什么是 Spring 中的控制反转(IoC)和依赖注入(DI)？为什么它们对应用程序的设计和可维护性有益？

2. 叙述 AOP(面向切面编程)的概念和主要用途。示例说明如何在 Spring 中使用 AOP。

### 五、实操题

在 Spring 中，使用 IoC 和 AOP 构建一个简单的图书管理系统，具体内容如下。

(1) 创建一个 Book 类，包含属性 id、title、author。

(2) 创建一个 BookService 接口及其实现类 BookServiceImpl，包含方法 addBook(Book book)、deleteBook(int id)。

(3) 使用 IoC，将 BookDao 注入到 BookServiceImpl 中。

(4) 在 BookServiceImpl 中，使用 AOP 添加环绕通知，记录每次添加和删除图书的操作时间。

要求：使用注解进行 Spring 的配置；书写相应的 JUnit 测试用例，测试 IoC 和 AOP 的配置是否正确。

### 开源精神与集体主义

Spring 框架是开源的，这体现了全球软件开发社群中的集体主义和合作精神。通过学习和使用 Spring，我们不仅提升了自己的技术能力，还应将这种开源和合作的精神应用到更广泛的社会生活中。

# 第 5 章
# SpringMVC 上手开发

学习要点

1. 理解 SpringMVC 的基本结构和工作流程。
2. 掌握基础的请求处理,包括 URL 映射、参数绑定等。
3. 学习如何使用 JSON 进行数据交换。
4. 熟悉如何使用 Postman 工具进行接口测试。

学习要点

1. SpringMVC 基础。
2. SpringMVC 的请求处理。
3. 数据交换和 Postman 接口测试。

本章知识点结构如图 5-1 所示。

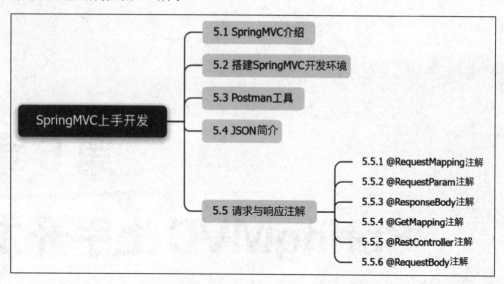

图 5-1　SpringMVC 上手开发

## 5.1　SpringMVC 介绍

动画：SpringMVC
——构建高效
Web 应用

　　SpringMVC(Model-View-Controller)是一个建立在 Spring 框架上的 Web 应用程序开发框架。它是 Spring Web MVC 的核心组件，提供了丰富的功能，包括数据绑定、Bean 验证和主题解析。SpringMVC 具有良好的分离性和可扩展性，允许开发者更加灵活和高效地构建复杂的 Web 应用程序。

　　SpringMVC 主要包括以下几个特点。

**1. 清晰的角色分离**

Controller(控制器)、Model(模型)和 View(视图)有明确的职责。

**2. 灵活的路由**

通过注解或 XML 配置，灵活地映射 URL 到具体的处理方法。

**3. 自动的数据绑定与验证**

自动的数据绑定、数据验证和错误处理。

**4. 多种视图渲染**

支持多种视图渲染技术，如 JSP、Thymeleaf、Freemarker 等。

**5. 扩展性**

通过插拔不同的组件和服务，可以轻松地扩展其功能。可与 Spring 生态系统集成，能无缝集成 Spring Security、Spring Data 等其他 Spring 项目。

## 5.2 搭建 SpringMVC 开发环境

微课：SpringMVC
开发环境搭建

下面通过使用 Maven 创建一个 SpringMVC 项目，来介绍 SpringMVC 的功能。

(1) 在 IntelliJ IDEA 中创建一个新的 Maven 项目 chapter05，添加相关依赖以及指定 packaging 打包方式为 war，代码如下：

```xml
<?xml version="1.0" encoding="UTF-8"?>
<project xmlns="http://maven.apache.org/POM/4.0.0"
     xmlns:xsi="http://www.w3.org/2001/XMLSchema-instance"
     xsi:schemaLocation="http://maven.apache.org/POM/4.0.0
         http://maven.apache.org/xsd/maven-4.0.0.xsd">
   <modelVersion>4.0.0</modelVersion>

   <groupId>com.demo</groupId>
   <artifactId>chapter05</artifactId>
   <version>1.0-SNAPSHOT</version>
   <!-- 设置打包方式 -->
   <packaging>war</packaging>

   <properties>
      <maven.compiler.source>8</maven.compiler.source>
      <maven.compiler.target>8</maven.compiler.target>
      <project.build.sourceEncoding>UTF-8</project.build.sourceEncoding>
   </properties>

   <dependencies>
      <!-- Servlet 包 -->
      <dependency>
         <groupId>javax.servlet</groupId>
         <artifactId>javax.servlet-api</artifactId>
         <version>4.0.1</version>
         <scope>provided</scope>
      </dependency>

      <!-- 导入 SpringMVC 的包 -->
      <dependency>
         <groupId>org.springframework</groupId>
         <artifactId>spring-webmvc</artifactId>
         <version>5.2.10.RELEASE</version>
      </dependency>

      <!-- JSON 转换 -->
      <dependency>
         <groupId>com.fasterxml.jackson.core</groupId>
         <artifactId>jackson-databind</artifactId>
         <version>2.9.0</version>
      </dependency>
   </dependencies>

</project>
```

(2)在项目的 src/main/java 目录下创建 com.demo.config 包，在该包下创建 ServletConfigInitializer 类，用于初始化 Web 容器。代码如下：

```java
package com.demo.config;

import org.springframework.web.context.WebApplicationContext;
import org.springframework.web.context.support
    .AnnotationConfigWebApplicationContext;
import org.springframework.web.filter.CharacterEncodingFilter;
import org.springframework.web.servlet.support
    .AbstractDispatcherServletInitializer;

import javax.servlet.Filter;

//这里定义我们的 ServletConfigInitializer 类
//继承了 AbstractDispatcherServletInitializer
//这样做的好处是，它提供了一种简洁的方式来初始化 DispatcherServlet 和 Spring 应用上下文
public class ServletConfigInitializer
    extends AbstractDispatcherServletInitializer {

    //此方法用于创建一个 Servlet 级别的应用上下文
    //在这个方法里，我们创建了一个 AnnotationConfigWebApplicationContext 对象
    //并通过 register()方法注册了一个配置类 SpringMvcConfig
    protected WebApplicationContext createServletApplicationContext() {
        //初始化 WebApplicationContext 对象
        AnnotationConfigWebApplicationContext ctx =
            new AnnotationConfigWebApplicationContext();
        //加载指定配置类
        ctx.register(SpringMvcConfig.class);
        return ctx;
    }

    //这个方法返回一个字符串数组,用于指定哪些路径应由 DispatcherServlet 处理
    //在这里,我们指定了所有请求("/")都应由 DispatcherServlet 处理
    protected String[] getServletMappings() {
        return new String[]{"/"};
    }

    //这个方法用于创建一个 Root 级别的应用上下文
    //在这个例子中，我们返回 null,因为我们只关心 Servlet 级别的应用上下文
    protected WebApplicationContext createRootApplicationContext() {
        return null;
    }

    //这个方法返回一个 Filter 数组,用于指定哪些 Filter 应用是由 DispatcherServlet
    //处理的请求
    // 在这个例子中,我们设置了一个 CharacterEncodingFilter,并设置其字符集为 utf-8
    @Override
    protected Filter[] getServletFilters() {
        CharacterEncodingFilter characterEncodingFilter =
            new CharacterEncodingFilter();
        characterEncodingFilter.setEncoding("utf-8");
        return new Filter[]{characterEncodingFilter};
    }
}
```

(3) 在 com.demo.config 包下创建 SpringMvcConfig 类，代码如下：

```java
package com.demo.config;

import org.springframework.context.annotation.ComponentScan;
import org.springframework.context.annotation.Configuration;

// SpringMVC 主配置类
@Configuration
@ComponentScan("com.demo.controller")
@EnableWebMvc
public class SpringMvcConfig {
}
```

(4) 在项目的 src/main/java 目录下创建 com.demo.controller 包，在该包下创建 BookController 类。代码如下：

```java
package com.demo.controller;

import org.springframework.stereotype.Controller;
import org.springframework.web.bind.annotation.RequestMapping;
import org.springframework.web.bind.annotation.ResponseBody;

@Controller
public class BookController {

    @RequestMapping("/add")
    @ResponseBody
    public String add(){
        System.out.println("BookController 的 add 方法执行了");
        return "springmvc ok";
    }

}
```

(5) IntelliJ IDEA 集成 Tomcat。

① 下载 Tomcat，本书中使用版本为 8.5.34，访问地址 https://archive.apache.org/dist/tomcat/tomcat-8/v8.5.34/bin/apache-tomcat-8.5.34.zip 下载 Tomcat 安装包。

② 将下载的 apache-tomcat-8.5.34.zip 解压缩到 D 盘，如图 5-2 所示。

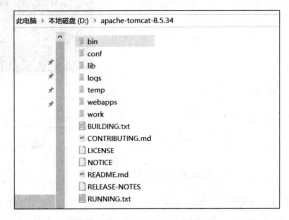

图 5-2  Tomcat 解压目录

③ 在 IntelliJ IDEA 中安装 Tomcat 插件，选择 File→Settings 命令，如图 5-3 所示。

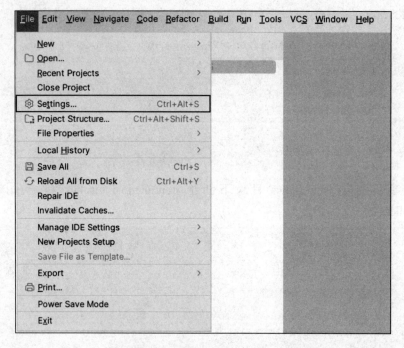

图 5-3 选择 Settings 命令

④ 在左侧列表中选择 Plugins，在右侧输入框搜索 Tomcat，安装 Smart Tomcat 插件，插件安装成功后重启 IDEA，如图 5-4 所示。

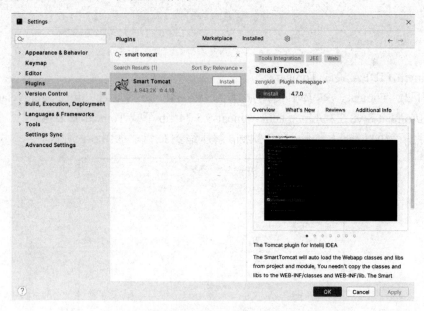

图 5-4 安装 Tomcat 插件

⑤ 重启之后，在右上角选择 Edit Configurations 选项，如图 5-5 所示。

⑥ 单击左上角"+"图标，添加 Smart Tomcat，如图 5-6 所示。

图 5-5 配置 Tomcat

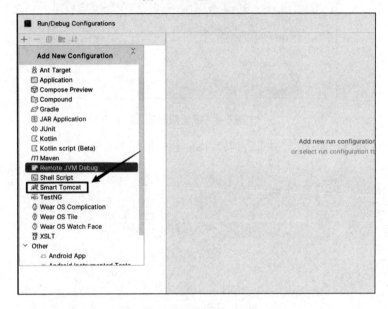

图 5-6 添加 Smart Tomcat

⑦ 输入 Tomcat 相关的配置信息，如图 5-7、图 5-8 所示。

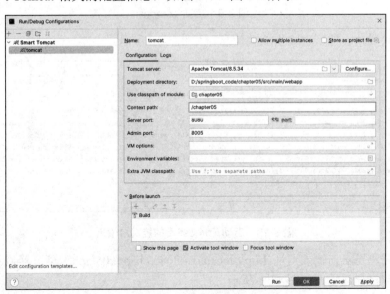

图 5-7 输入配置信息

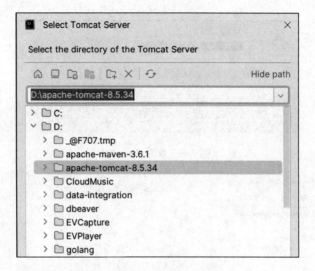

图 5-8 配置服务器

⑧ 配置成功后,单击 ▷ 按钮,启动 Tomcat,如图 5-9 所示。

图 5-9 启动 Tomcat

⑨ 启动成功,控制台输出相关信息,如图 5-10 所示。

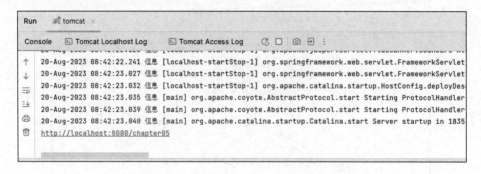

图 5-10 启动 Tomcat 后的控制台输出

(6) 打开浏览器,输入网址 http://localhost:8080/chapter05/add。浏览器显示 springmvc ok,说明环境搭建成功,如图 5-11 所示。

图 5-11　访问项目

## 5.3　Postman 工具

Postman 是一款功能强大的发送 HTTP 请求的测试工具，使用它 Web 开发测试人员可以创建和发送各种 HTTP 请求以完成 HTTP 接口的测试工作。在本书后面的章节中，将使用 Postman 工具来测试所开发的 HTTP 接口。

下面简要介绍 Postman 的安装和使用流程。

（1）访问链接 https://www.postman.com/downloads/，选择对应的版本下载。本书使用 7.9.0 版本。下载 Postman 软件安装包，双击安装文件，如图 5-12 所示。

动画：Postman 探索——您的 API 开发利器　　微课：Postman 工具应用

图 5-12　Postman 安装包

（2）跳过登录，直接进入 Postman 主界面，如图 5-13、图 5-14 所示。

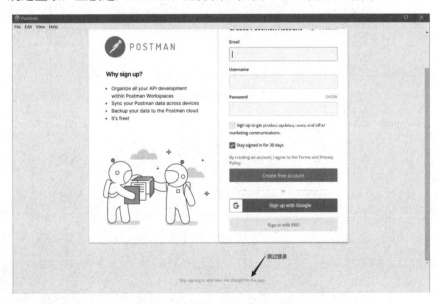

图 5-13　Postman 登录界面

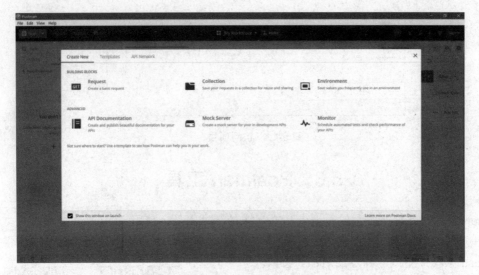

图 5-14 Postman 主界面

(3) 创建接口集合，如图 5-15～图 5-17 所示。

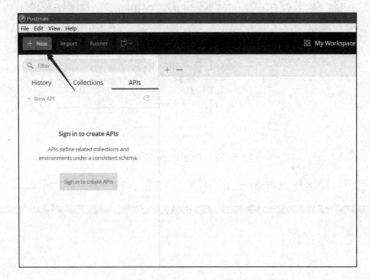

图 5-15 Postman 新建

图 5-16 新建接口集合

图 5-17　接口集合命名

(4) 在新建接口集合下创建请求，如图 5-18、图 5-19 所示。

图 5-18　新建请求接口

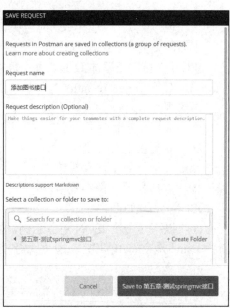

图 5-19　请求接口命名

(5) 在新建的请求接口下通过 GET 请求发送 url，单击 Send 按钮发送请求，如图 5-20 所示。

图 5-20　发送请求

## 5.4　JSON 简介

动画：JSON 揭秘
——数据交换的
简易之选

　　JSON(JavaScript Object Notation)是一种轻量级的数据交换格式，易于人工阅读和编写，同时也易于机器解析和生成。尽管 JSON 是 JavaScript 的一个子集，但是它与语言无关，并且在几乎所有现代编程语言中都有对应的库来处理 JSON 数据。在现今绝大多数的 Web 应用中，HTTP 接口都以 JSON 数据作为返回结果格式。

　　下面简要介绍 JSON 格式。

　　JSON 支持以下基本数据类型：数字(Number)、字符串(String)、布尔值(Boolean)、数组(Array)、对象(Object)、null。

　　语法规则：

◎　数据由键值对构成。
◎　数据由逗号分隔。
◎　大括号用于保存对象。
◎　方括号用于保存数组。

JSON 示例如下。

(1) 以下是一个 JSON 对象示例，对象中包含三个属性，分别是 name、age 和 is_student，其中 name 是字符串类型，age 是数字类型，is_student 是布尔型：

```
{
  "name": "John",
  "age": 30,
  "is_student": false
}
```

(2) 以下是一个 JSON 数组示例，数组长度为 2，包含了两个 JSON 对象：

```
[
  {
```

```
    "name": "John",
    "age": 30,
    "is_student": false
  },
  {
    "name": "Jane",
    "age": 22,
    "is_student": true
  }
]
```

## 5.5 请求与响应注解

### 5.5.1 @RequestMapping 注解

在 SpringMVC 和 SpringBoot 应用程序中，@RequestMapping 注解用于映射 Web 请求(例如 HTTP 请求)到特定的处理器方法或者类上。这个注解是 SpringMVC 框架中最常用的一个。

动画：深度解析 SpringMVC——优雅地处理请求与响应

微课：SpringMVC 请求响应详解

@RequestMapping 注解有以下两种使用方式。

(1) 映射到方法：使用@RequestMapping 注解，可以指定一个或多个 HTTP 方法(如 GET、POST、PUT、DELETE 等)以及请求路径。

value(或 path)：指定请求的实际 URL，例如@RequestMapping("/hello")。

method：指定 HTTP 请求的类型，如 GET、POST 等。

(2) 映射到类：将@RequestMapping 用在控制器类上，这样类中的所有方法都会有一个共同的路径前缀。

下面通过一个简单示例，演示@RequestMapping 映射到方法注解的使用。

(1) 在 com.demo.controller 包下新建一个 ApiController 控制器，代码如下：

```
package com.demo.controller;

import org.springframework.stereotype.Controller;
import org.springframework.web.bind.annotation.RequestMapping;
import org.springframework.web.bind.annotation.RequestMethod;
import org.springframework.web.bind.annotation.ResponseBody;

@Controller
public class ApiController {

    @RequestMapping(value = "/hello", method = RequestMethod.GET)
    @ResponseBody
    public String sayHello() {
        return "Hello, World!";
    }
}
```

(2) 通过 Postman 工具对上述接口进行测试，创建一个新的请求接口，如图 5-21 所示。
(3) 输入接口的请求地址 http://localhost:8080/chapter05/hello，单击 Send 按钮，如图 5-22 所示。

图 5-21　新建请求接口

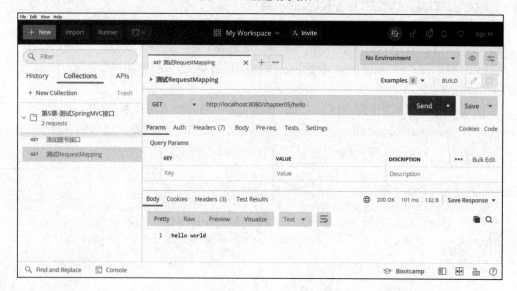

图 5-22　接口测试

下面通过一个简单示例，演示@RequestMapping 映射到类注解的使用。
(1) 修改前面的代码如下：

```
@Controller
@RequestMapping("/api")
public class ApiController {
```

```
@RequestMapping(value = "/hello", method = RequestMethod.GET)
@ResponseBody
public String sayHello() {
    return "Hello, World!";
}
```

(2) 输入接口的请求地址 http://localhost:8080/chapter05/api/hello，单击 Send 按钮，如图 5-23 所示。

图 5-23  接口测试

## 5.5.2  @RequestParam 注解

在 SpringMVC 和 SpringBoot 应用程序中，@RequestParam 注解用于从请求的查询字符串部分获取参数值。这个注解通常被用在 SpringMVC 的控制器(Controller)中。

name/value：指定请求参数的名字。

defaultValue：如果请求中没有指定参数，将使用默认值。

required：指定该参数是否必须。默认值是 true。

下面通过一个简单示例，演示@RequestParam 注解的使用。

(1) 在 ApiController 控制器中增加 greet()方法，代码如下：

```
@Controller
@RequestMapping("/api")
public class ApiController {

@RequestMapping("/greet")
@ResponseBody
    public String greet(@RequestParam(name = "name", defaultValue = "Guest")
        String name) {
        return "Hello, " + name + "!";
    }

}
```

(2) 在这个例子中，HTTP 请求 /api/greet?name=John 将输出"Hello, John!"。如果没有提供 name 参数(如 /api/greet)，则使用默认值 "Guest"，输出"Hello, Guest!"，如图 5-24、图 5-25 所示。

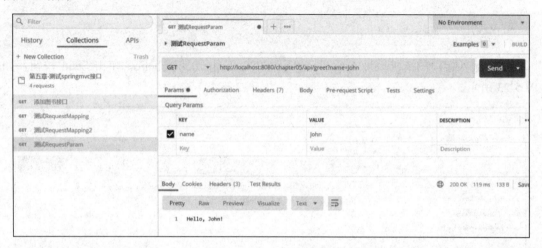

图 5-24　接口测试(带参数)

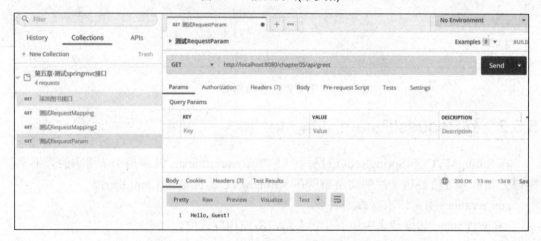

图 5-25　接口测试(不带参数)

(3) 在 ApiController 控制器中增加 search 方法，代码如下：

```java
@RequestMapping("/search")
@ResponseBody
public String search(@RequestParam("keyword") String keyword,
        @RequestParam(value = "sort", required = false) String sort) {
    return "Search results for keyword: " + keyword + " and sort: " + sort;
}
```

(4) 请求 /api/search?keyword=apple 将返回"Search results for keyword: apple and sort: null"。

请求/api/search?keyword=apple&sort=desc 将返回"Search results for keyword: apple and sort: desc"，如图 5-26、图 5-27 所示。

图 5-26　接口测试(单参数)

图 5-27　接口测试(多参数)

### 5.5.3 @ResponseBody 注解

在 SpringMVC 中，@ResponseBody 注解用于将 Controller 方法返回的对象通过适当的 HttpMessageConverter 转换为指定格式后，写入到 HTTP 响应(Response)体中。通常用于返回 JSON 或 XML 数据。

下面通过一个简单示例，演示 @ResponseBody 注解的使用。

(1) 在项目的 src/main/java 目录下创建 com.demo.pojo 包，然后创建一个商品实体类，代码如下：

```java
package com.demo.pojo;

public class Product {

    private int id;
    private String name;
    private double price;
```

```
    //省略无参数、有参数构造方法
    //省略属性的get()、set()方法
}
```

(2) 在ApiController类中增加getProduct方法，测试@ResponseBody注解，代码如下：

```
@RequestMapping("/getProduct")
@ResponseBody
public Product getProduct(@RequestParam("id") int id) {
    // 从数据库或其他地方获取商品信息
    return new Product(id, "Apple", 1.2);
}
```

(3) 输入接口的请求地址http://localhost:8080/chapter05/api/getProduct?id=1，单击Send按钮，如图5-28所示。

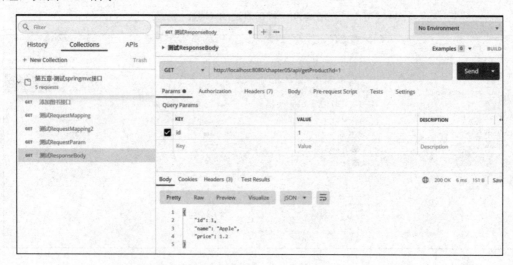

图5-28 接口测试

### 5.5.4 @GetMapping注解

@GetMapping是Spring Web模块中的一个注解，用于处理HTTP GET请求。它是@RequestMapping注解的一个特化版本，专门用来处理GET请求类型。这个注解通常与Controller方法一起使用，来指定该方法应该响应哪种类型的HTTP GET请求。像这样的注解还有@DeleteMapping、@PostMapping、@PutMapping等，分别用于处理增删改查操作。

下面通过一个简单示例，演示@GetMapping注解的使用。

(1) 在ApiController控制器中增加welcome()方法，代码如下：

```
@GetMapping("/welcome")
@ResponseBody
public String welcome() {
    return "Hello, World!";
}
```

(2) 在这个例子中，当访问http://localhost:8080/welcome时，sayHello方法将被执行，并返回字符串"Hello, World!"。

## 5.5.5 @RestController 注解

@RestController 是 Spring Framework 中一个非常重要的注解,用于定义一个 RESTful Web Service 的 Controller 类。这个注解是 @Controller 和 @ResponseBody 两个注解的组合,表明这是一个用于处理 HTTP 请求并返回 JSON 或 XML 响应体的 Controller。

若标注一个类为 @RestController,这个类中所有的方法都会默认添加 @ResponseBody 注解,这样方法返回的对象会自动转换为 JSON 或 XML 格式并返回给客户端。例如:

```
import org.springframework.web.bind.annotation.GetMapping;
import org.springframework.web.bind.annotation.RestController;

@RestController
public class HelloController {

    @GetMapping("/hello")
    public String sayHello() {
        return "Hello, World!";
    }
}
```

这个例子中,HelloController 类标注了 @RestController 注解,因此它会处理 HTTP 请求并返回响应。sayHello()方法通过 @GetMapping 注解映射到 /hello URL,并返回一个简单的"Hello, World!"字符串。

## 5.5.6 @RequestBody 注解

@RequestBody 是 Spring Framework 中用于处理 HTTP 请求体的注解。当客户端向服务器发送一个请求时,除了 URL 和头信息外,还可以附带一个请求体。@RequestBody 注解告诉 Spring 将 HTTP 请求体转换并绑定到方法参数上。这通常用于解析 JSON 或 XML 格式的数据。

下面是一个使用 @RequestBody 的简单示例,演示了如何在 SpringBoot 应用中接收 JSON 格式的数据。

(1) 创建示例,代码如下:

```
@RestController
@RequestMapping("/api")
public class ApiController {

    @PostMapping("createProduct")
    public Product createProduct(@RequestBody Product product){
        // 在这里可以对 user 对象进行各种操作,例如保存到数据库
        return product;
    }
}
```

(2) 使用 Postman 测试接口,如图 5-29 所示。

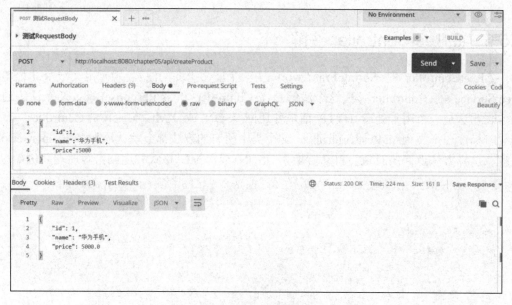

图 5-29　接口测试

## 本 章 小 结

本章主要介绍了 SpringMVC 框架，这是构建 Web 应用程序的一个重要部分。我们学习了如何搭建 SpringMVC 开发环境，理解了基本的请求处理和数据交换格式；通过 Postman 工具，也进行了实际的接口测试，以更好地理解 SpringMVC 的工作方式。

## 课 后 习 题

一、选择题

1. SpringMVC 是 Spring 框架的一部分，它主要用于(　　)。
　　A. 数据库管理　　　　　　　　B. 控制反转
　　C. Web 应用程序开发　　　　　D. 科学计算
2. 在 SpringMVC 中，请求处理方法应该使用(　　)注解来标记。
　　A. @Controller　　B. @Service　　C. @Repository　　D. @Component
3. 在 SpringMVC 中，用于将 Java 对象转换为 JSON 格式的默认库是(　　)。
　　A. Gson　　　　　B. Jackson　　　C. FastJSON　　　D. XML

二、填空题

1. 要创建一个 SpringMVC 环境，通常需要在项目的配置文件中配置一个_____。
2. Postman 是一个流行的工具，用于进行_____测试和调试。

三、判断题

1. SpringMVC 仅适用于基于 Java 的 Web 应用程序开发。　　　　　　　　(　　)

2. 在 SpringMVC 中，我们可以使用多种方法来处理和生成 JSON 数据，而不一定非要使用 Jackson 库。（　　）

### 四、简答题

1. 如何使用 Postman 进行 API 测试和调试？请提供一个示例，说明如何发送 GET 和 POST 请求以及验证响应。

2. 在 SpringMVC 中，如何处理 JSON 数据？请提供一个示例，说明如何将 Java 对象转换为 JSON 格式，并将 JSON 响应发送到客户端。

### 五、实操题

使用 SpringMVC 构建一个简单的用户管理系统，内容如下。

(1) 创建一个 User 类，包含属性 id、name、email。

(2) 创建一个 UserService 接口及其实现类 UserServiceImpl，包含方法 addUser(User user)、deleteUser(int id)、getAllUsers()。

(3) 创建一个名为 UserController 的控制器，里面应包含以下方法：添加用户、删除用户、列出所有用户。

(4) 使用 SpringMVC 注解(如@RequestMapping、@RequestParam 等)配置 UserController。

要求：使用注解进行 SpringMVC 的配置；使用 Postman 进行接口测试。

服务端与客户端，像政府与公民

SpringMVC 常用于构建具有复杂业务逻辑的企业级应用，起着服务端的作用，处理来自客户端(用户)的各种请求。这种模型与现实社会中政府与公民的关系类似。政府需要对公民的需求做出响应，合理分配资源，制定并执行政策。作为服务端的开发人员，我们也需要以用户为中心，高效、公平地处理请求，同时确保安全和合规性。这样的角度不仅能帮助我们更好地理解服务端开发的核心任务，也能让我们在开发过程中始终牢记社会责任和公平原则。

# 第 6 章
# SSM 整合开发

**学习目标**

1. 掌握如何配置 Spring 与 MyBatis,以及 SpringMVC 与 Spring 的整合。
2. 学会在 SSM 架构中实现 CRUD(增删查改)操作。

**学习要点**

1. SSM 框架整合需要的配置。
2. MyBatis 的 Mapper 接口和映射文件。
3. 基于 Postman 对接口进行测试。

本章知识点结构如图 6-1 所示。

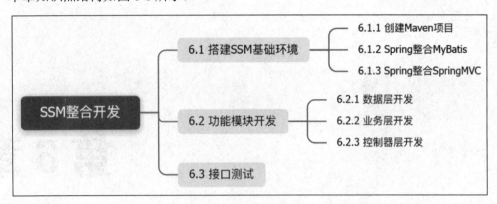

图 6-1　SSM 整合开发

## 6.1　搭建 SSM 基础环境

在前面的章节中，本书已经介绍了 MyBatis、Spring 和 SpringMVC 这三个 Java 语言开发的利器。在实际的 Java Web 开发中，通常是将这三者整合起来，通过 Spring 的强大灵活的对象管理机制整合 MyBatis 和 SpringMVC，实现 Java Web 应用的高效开发。

下面将通过一个简单的图书信息管理程序来演示整合的过程。

动画：SSM 的整合　　微课：搭建 SSM 基础环境

### 6.1.1　创建 Maven 项目

在 IntelliJ IDEA 中创建一个新的 Maven 项目，项目名为 chapter06，在 pom.xml 文件中添加相关依赖，代码如下：

```xml
<?xml version="1.0" encoding="UTF-8"?>
<project xmlns="http://maven.apache.org/POM/4.0.0"
     xmlns:xsi="http://www.w3.org/2001/XMLSchema-instance"
     xsi:schemaLocation="http://maven.apache.org/POM/4.0.0
        http://maven.apache.org/xsd/maven-4.0.0.xsd">
    <modelVersion>4.0.0</modelVersion>

    <groupId>com.demo</groupId>
    <artifactId>chapter06</artifactId>
    <version>1.0-SNAPSHOT</version>
    <packaging>war</packaging>

    <properties>
        <maven.compiler.source>8</maven.compiler.source>
        <maven.compiler.target>8</maven.compiler.target>
        <project.build.sourceEncoding>UTF-8</project.build.sourceEncoding>
        <spring.version>5.2.10.RELEASE</spring.version>
    </properties>
```

```xml
<dependencies>
<!-- Spring 依赖 -->
    <dependency>
        <groupId>org.springframework</groupId>
        <artifactId>spring-webmvc</artifactId>
        <version>${spring.version}</version>
    </dependency>

    <dependency>
        <groupId>org.springframework</groupId>
        <artifactId>spring-jdbc</artifactId>
        <version>${spring.version}</version>
    </dependency>

    <dependency>
        <groupId>org.springframework</groupId>
        <artifactId>spring-test</artifactId>
        <version>${spring.version}</version>
    </dependency>

    <!-- MyBatis 依赖 -->
    <dependency>
        <groupId>org.mybatis</groupId>
        <artifactId>mybatis</artifactId>
        <version>3.5.6</version>
    </dependency>

    <!-- Spring 整合 MyBatis 依赖 -->
    <dependency>
        <groupId>org.mybatis</groupId>
        <artifactId>mybatis-spring</artifactId>
        <version>1.3.0</version>
    </dependency>

    <!-- MySQL 数据库驱动依赖 -->
    <dependency>
        <groupId>mysql</groupId>
        <artifactId>mysql-connector-java</artifactId>
        <version>5.1.17</version>
    </dependency>

    <!-- 数据库连接池依赖 -->
    <dependency>
        <groupId>com.alibaba</groupId>
        <artifactId>druid</artifactId>
        <version>1.1.16</version>
    </dependency>

    <!-- 单元测试依赖 -->
    <dependency>
        <groupId>junit</groupId>
        <artifactId>junit</artifactId>
```

```xml
            <version>4.12</version>
            <scope>test</scope>
        </dependency>

        <!-- Servlet 依赖 -->
        <dependency>
            <groupId>javax.servlet</groupId>
            <artifactId>javax.servlet-api</artifactId>
            <version>3.1.0</version>
            <scope>provided</scope>
        </dependency>

        <!-- Lombok 依赖 -->
        <dependency>
            <groupId>org.projectlombok</groupId>
            <artifactId>lombok</artifactId>
            <version>1.18.16</version>
            <scope>provided</scope>
        </dependency>

        <!-- Jackson 依赖 -->
        <dependency>
            <groupId>com.fasterxml.jackson.core</groupId>
            <artifactId>jackson-databind</artifactId>
            <version>2.9.0</version>
        </dependency>
    </dependencies>
</project>
```

## 6.1.2 Spring 整合 MyBatis

Spring 整合 Mybatis，即由 Spring 来管理 MyBatis 程序中的各项配置和对象的加载与创建，如数据库配置和 SqlSession 对象等。

(1) 在 MySQL 数据库下，创建数据库 ssm_db、创建图书信息表 book，表中包含四个字段，即图书编号 id、图书名称 name、图书类型 type 和图书描述 description。创建数据库和数据表的 SQL 语句如下：

```sql
-- 创建 ssm_db 数据库
CREATE DATABASE IF NOT EXISTS ssm_db CHARACTER SET utf8;

-- 使用 ssm_db 数据库
USE ssm_db;

-- 创建 tbl_book 表
CREATE TABLE t_book(
    id INT PRIMARY KEY AUTO_INCREMENT,  -- 图书编号
    `name` VARCHAR(100),                -- 图书名称
    `type` VARCHAR(100),                -- 图书类型
    description VARCHAR(100)            -- 图书描述
);
```

数据表创建好后，向 book 表中插入一些测试数据，SQL 语句如下：

```
INSERT INTO t_book (`name`, `type`, description) VALUES
('三体', '科幻', '描述文明的崩溃和重生'),
('红楼梦', '古典文学', '一部描写封建家庭悲剧的小说'),
('西游记', '古典文学', '描述孙悟空等人西行取经的故事'),
('水浒传', '古典文学', '描述梁山好汉反抗封建统治的故事'),
('哈利·波特', '奇幻', '描写一名年轻巫师的成长故事'),
('百年孤独', '现代文学', '描述布恩迪亚家族几代人的传奇故事'),
('围城', '现代文学', '描述现代都市生活和爱情的小说');
```

（2）在项目的 resources 目录下创建 jdbc.properties 文件，添加 MySQL 数据库的配置，主要是 JDBC 驱动、数据库地址、数据库用户名和密码，配置内容如下：

```
jdbc.driverClassName=com.mysql.jdbc.Driver
jdbc.url=jdbc:mysql://localhost:3306/ssm_db?useSSL=false&characterEncoding=utf8
jdbc.username=root
jdbc.password=123456
```

（3）在项目的 src/main/java 目录下，创建 com.demo.config 包。创建 JdbcConfig 配置类，配置数据库连接池、事务管理器，代码如下：

```java
package com.demo.config;

import com.alibaba.druid.pool.DruidDataSource;
import org.springframework.beans.factory.annotation.Value;
import org.springframework.context.annotation.Bean;
import org.springframework.context.annotation.PropertySource;
import org.springframework.jdbc.datasource.DataSourceTransactionManager;
import org.springframework.transaction.PlatformTransactionManager;
import javax.sql.DataSource;

@PropertySource("classpath:jdbc.properties")//加载外部配置文件jdbc.properties
public class JdbcConfig {

    //@Value 注解：获取 jdbc.properties 配置参数值，并赋值给当前属性
    @Value("${jdbc.driverClassName}")
    private String driver;
    @Value("${jdbc.url}")
    private String url;
    @Value("${jdbc.username}")
    private String username;
    @Value("${jdbc.password}")
    private String password;

    //配置数据库连接池
    @Bean
    public DataSource dataSource(){
        DruidDataSource dataSource = new DruidDataSource();
        dataSource.setDriverClassName(driver);
        dataSource.setUrl(url);
```

```
        dataSource.setUsername(username);
        dataSource.setPassword(password);
        return dataSource;
    }

    //Spring事务管理需要的平台事务管理器对象
    @Bean
    public PlatformTransactionManager transactionManager(DataSource dataSource){
        DataSourceTransactionManager ds = new DataSourceTransactionManager();
        ds.setDataSource(dataSource);
        return ds;
    }
}
```

(4) 在 com.demo.config 包下创建 MybatisConfig 配置类，该类主要用于加载 MyBatis 相关的信息，代码如下：

```
package com.demo.config;

import org.apache.ibatis.logging.stdout.StdOutImpl;
import org.apache.ibatis.session.Configuration;
import org.mybatis.spring.SqlSessionFactoryBean;
import org.mybatis.spring.mapper.MapperScannerConfigurer;
import org.springframework.context.annotation.Bean;
import org.springframework.core.io.Resource;
import org.springframework.core.io.support.PathMatchingResourcePatternResolver;
import org.springframework.core.io.support.ResourcePatternResolver;
import javax.sql.DataSource;
import java.io.IOException;

public class MybatisConfig {

    @Bean
    public SqlSessionFactoryBean sqlSessionFactoryBean(DataSource dataSource)
      throws IOException {
        SqlSessionFactoryBean sqlSessionFactoryBean = new SqlSessionFactoryBean();
        sqlSessionFactoryBean.setTypeAliasesPackage("com.demo.pojo");
        sqlSessionFactoryBean.setDataSource(dataSource);

        Configuration configuration = new Configuration();
        //配置将数据库表中的下划线格式的列名自动映射为驼峰格式的Java变量名
        configuration.setMapUnderscoreToCamelCase(true);
        //设置打印日志
        configuration.setLogImpl(StdOutImpl.class);
        sqlSessionFactoryBean.setConfiguration(configuration);

        //加载resources下的MyBatis配置文件
        ResourcePatternResolver resolver = new PathMatchingResourcePatternResolver();
        Resource[] resources = resolver.getResources("classpath*:mappers/
          *Mapper.xml");
        sqlSessionFactoryBean.setMapperLocations(resources);
```

```java
        return sqlSessionFactoryBean;
    }

    @Bean
    public MapperScannerConfigurer mapperScannerConfigurer(){
        MapperScannerConfigurer mapperScannerConfigurer =
          new MapperScannerConfigurer();
            //设置MyBatis加载映射接口位置
        mapperScannerConfigurer.setBasePackage("com.demo.mapper");
        return mapperScannerConfigurer;
    }
}
```

(5) 在 com.demo.config 包下创建 SpringConfig 配置类, 代码如下:

```java
package com.demo.config;

import org.springframework.context.annotation.*;
import org.springframework.stereotype.Controller;
import org.springframework.transaction.annotation.EnableTransactionManagement;

//扫描 com.demo 下除了 Controller 注解以外的配置类
@ComponentScan(value = "com.demo",
    excludeFilters = @ComponentScan.Filter(
        type = FilterType.ANNOTATION,
        classes = Controller.class
    )
)
//导入配置类
@Import({JdbcConfig.class,MybatisConfig.class})
//启用事务管理功能
@EnableTransactionManagement
public class SpringConfig {
}
```

## 6.1.3 Spring 整合 SpringMVC

Spring 整合 SpringMVC, 即使用 Spring 的对象管理功能管理 SpringMVC 相关的对象。其详细开发步骤如下。

(1) 在 com.demo.config 包下创建 SpringMvcConfig 配置类, 代码如下:

```java
package com.demo.config;

import org.springframework.context.annotation.ComponentScan;
import org.springframework.web.servlet.config.annotation.EnableWebMvc;
import org.springframework.web.servlet.config.annotation.WebMvcConfigurer;

//扫描 com.demo.controller 下的配置类, 主要是控制器
```

```
@ComponentScan("com.demo.controller")
//启用SpringMVC
@EnableWebMvc
public class SpringMvcConfig implements WebMvcConfigurer {

}
```

(2) 在 com.demo.config 包下创建 ServletConfigInitializer 配置类，代码如下：

```
package com.demo.config;

import org.springframework.web.context.WebApplicationContext;
import org.springframework.web.context.support
    .AnnotationConfigWebApplicationContext;
import org.springframework.web.filter.CharacterEncodingFilter;
import org.springframework.web.servlet.support
    .AbstractDispatcherServletInitializer;

import javax.servlet.Filter;

public class ServletConfig extends AbstractDispatcherServletInitializer {

    //加载SpringMvcConfig配置类
    @Override
    protected WebApplicationContext createServletApplicationContext() {
        AnnotationConfigWebApplicationContext ac =
          new AnnotationConfigWebApplicationContext();
        ac.register(SpringMvcConfig.class);
        return ac;
    }

  //拦截URL请求，交给SpringMVC处理
    @Override
    protected String[] getServletMappings() {
        return new String[]{"/"};
    }

    //加载SpringConfig配置类
    @Override
    protected WebApplicationContext createRootApplicationContext() {
        AnnotationConfigWebApplicationContext ac =
            new AnnotationConfigWebApplicationContext();
        ac.register(SpringConfig.class);
        return ac;
    }

    //配置SpringMVC框架提供的处理Post请求中文乱码过滤器
    @Override
    protected Filter[] getServletFilters() {
```

```
        CharacterEncodingFilter characterEncodingFilter =
            new CharacterEncodingFilter();
        characterEncodingFilter.setEncoding("utf-8");
        return new Filter[]{characterEncodingFilter};
    }
}
```

## 6.2 功能模块开发

### 6.2.1 数据层开发

微课：功能模块开发

数据层即与数据库访问相关的代码，主要包括与数据表对应的 Java Bean 和 MyBatis 相关的数据处理类等。详细开发步骤如下。

（1）在项目的 src/main/java 目录下，创建 com.demo.pojo 包，在该包下创建 Book 类，代码如下：

```
package com.demo.pojo;

import lombok.Data;
//@Data 注解加在类上可以在程序编译期间自动生成各字段的 getter 和 setter 等方法使程序简洁
@Data
public class Book {

    private Integer id;
    private String type;
    private String name;
    private String description;

}
```

（2）在项目的 src/main/java 目录下，创建 com.demo.mapper 包。在该包下创建 BookMapper 接口，代码中包含对图书信息的添加、更新、删除，根据 id 查询及查询的所有结果，代码如下：

```
package com.demo.mapper;

import com.demo.pojo.Book;

import java.util.List;

public interface BookMapper {
    //添加 Book 信息
    public int save(Book book);
    //修改 Book 信息
    public int update(Book book);
    //删除 Book 信息
    public int delete(Integer id);
    //根据 id 查询 Book 信息
    public Book getById(Integer id);
```

```java
//查询全部Book信息
public List<Book> getAll();
}
```

(3) 在 resources 目录下创建 mappers 目录，并且添加 BookMapper.xml 文件，在 XML 文件中添加与 BookMapper 接口中方法相对应的 MyBatis 增删改查的标签，代码如下：

```xml
<?xml version="1.0" encoding="UTF-8"?>
<!DOCTYPE mapper
 PUBLIC "-//mybatis.org//DTD Mapper 3.0//EN"
 "http://mybatis.org/dtd/mybatis-3-mapper.dtd">

<!-- namespace 应该由包名+接口名构成 -->
<mapper namespace="com.demo.mapper.BookMapper">

    <!-- id应该和BookMapper接口的方法名一致 -->
    <insert id="save">
        insert into t_book values(null,#{name},#{type},#{description})
    </insert>

    <update id="update">
        update t_book
        <set>
            <if test="name != null">
                name = #{name},
            </if>
            <if test="type != null">
                type = #{type},
            </if>
            <if test="description != null">
                description = #{description},
            </if>
        </set>
        where id = #{id}
    </update>

    <delete id="delete">
        delete from t_book where id = #{id}
    </delete>

    <select id="getById" resultType="com.demo.pojo.Book">
        select * from t_book where id = #{id}
    </select>

    <select id="getAll" resultType="com.demo.pojo.Book">
        select * from t_book
    </select>
</mapper>
```

## 6.2.2 业务层开发

业务层介于数据层和 Web 层之间，主要是在数据层的基础上完成系统的各个功能模块的具体实现，一般命名为 xxService，如本例中为 BookService，其详细开发步骤如下。

(1) 在项目的 src/main/java 目录下，创建 com.demo.service 包，在该包下创建 BookService 接口，代码如下：

```java
package com.demo.service;

import com.demo.pojo.Book;

import java.util.List;

public interface BookService {

    //添加 Book 信息
    public boolean save(Book book);

    //修改 Book 信息
    public boolean update(Book book);

    //删除 Book 信息
    public boolean delete(Integer id);

    //根据 id 查询 Book 信息
    public Book getById(Integer id);

    //查询全部 Book 信息
    public List<Book> getAll();
}
```

(2) 在项目的 src/main/java 目录下，创建 com.demo.service.impl 包，在该包下创建 BookService 接口的实现类 BookServiceImpl 类，代码如下：

```java
package com.demo.service.impl;

import com.demo.mapper.BookMapper;
import com.demo.pojo.Book;
import com.demo.service.BookService;
import org.springframework.beans.factory.annotation.Autowired;
import org.springframework.stereotype.Service;
import org.springframework.transaction.annotation.Transactional;

import java.util.List;

@Service
```

```
@Transactional
public class BookServiceImpl implements BookService {
    @Autowired
    private BookMapper bookDao;

    public boolean save(Book book) {
        bookDao.save(book);
        return true;
    }

    public boolean update(Book book) {
        bookDao.update(book);
        return true;
    }

    public boolean delete(Integer id) {
        bookDao.delete(id);
        return true;
    }

    public Book getById(Integer id) {
        return bookDao.getById(id);
    }

    public List<Book> getAll() {
        return bookDao.getAll();
    }
}
```

### 6.2.3 控制器层开发

控制器层即 SpringMVC 的 Controller 层，在业务层之上，完成对用户请求的处理和结果返回等功能。

在项目的 src/main/java 目录下，创建 com.demo.controller 包，在该包下创建 BookController 类。在对应的方法上可以添加 RequestMethod.POST、RequestMethod.PUT、RequestMethod.DELETE、RequestMethod.GET 四种注解用于处理不同的请求方式，分别用于处理增删改查操作。代码如下：

```
package com.demo.controller;

import com.demo.pojo.Book;
import com.demo.service.BookService;
```

```java
import org.springframework.beans.factory.annotation.Autowired;
import org.springframework.stereotype.Controller;
import org.springframework.web.bind.annotation.*;

import java.util.List;

@Controller
@RequestMapping("/books")
public class BookController {

    @Autowired
    private BookService bookService;

    @RequestMapping(method = RequestMethod.POST)
    @ResponseBody
    public boolean save(Book book) {
        return bookService.save(book);
    }

    @RequestMapping(method = RequestMethod.PUT)
    @ResponseBody
    public boolean update(Book book) {
        return bookService.update(book);
    }

    @RequestMapping(method = RequestMethod.DELETE)
    @ResponseBody
    public boolean delete(Integer id) {
        return bookService.delete(id);
    }

    @RequestMapping(method = RequestMethod.GET,value = "getById")
    @ResponseBody
    public Book getById(Integer id) {
        return bookService.getById(id);
    }

    @RequestMapping(method = RequestMethod.GET)
    @ResponseBody
    public List<Book> getAll() {
        return bookService.getAll();
    }
}
```

项目包与类结构如图 6-2 所示。

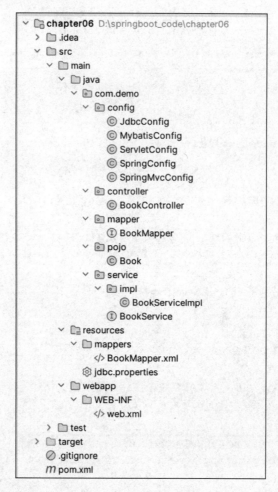

图 6-2　SSM 项目结构

## 6.3　接　口　测　试

微课：Postman
接口测试

在第 5 章中介绍了 Web 测试工具 Postman 的使用，本节继续使用 Postman 来完成对实例项目 Web 接口的测试，详细步骤如下。

（1）在 Postman 中针对上述开发接口进行测试，首先单击 New Collection，在打开的窗口中输入"第六章-SSM 整合"，如图 6-3 所示。

（2）在新创建的 Collection 下添加一个请求，如图 6-4、图 6-5 所示。

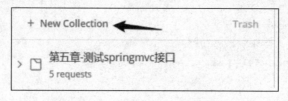

图 6-3　创建 Postman 集合

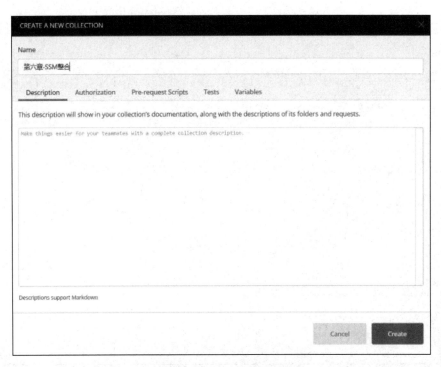

图 6-3 创建 Postman 集合(续)

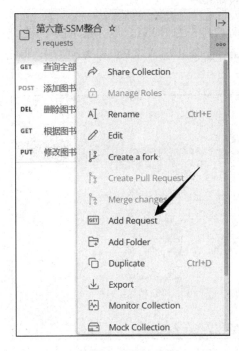

图 6-4 新建请求　　　　　　　　图 6-5 输入请求名称

(3) 测试查询全部图书信息的接口。在请求中选择请求方式为 GET，输入请求的 URL 地址 http://localhost:8080/chapter06/books，单击 Send 按钮，查看下方返回的 JSON 数据，如图 6-6 所示。

图 6-6 查询全部图书信息

(4) 测试添加图书信息的接口。在请求中选择请求方式为 POST，输入请求的 URL 地址 http://localhost:8080/chapter06/books，在下方 Params 参数中输入 KEY 和 VALUE。单击 Send 按钮，如图 6-7 所示。

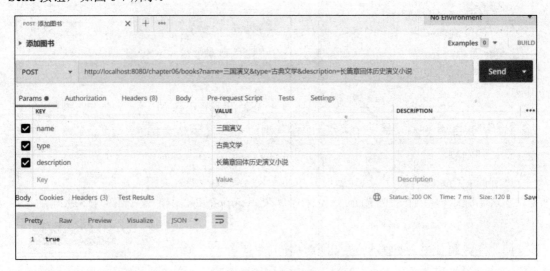

图 6-7 添加图书信息

(5) 测试根据图书编号查询图书信息的接口。在请求中选择请求方式为 GET，输入请求的 URL 地址 http://localhost:8080/chapter06/books/getById，在下方 Params 参数中输入 KEY 和 VALUE。单击 Send 按钮，如图 6-8 所示。

图 6-8　根据图书编号查询图书信息

（6）测试根据图书编号修改图书信息的接口。在请求中选择请求方式为 PUT，输入请求的 URL 地址 http://localhost:8080/chapter06/books，在下方 Params 参数中输入 KEY 和 VALUE。单击 Send 按钮，如图 6-9 所示。

图 6-9　根据图书编号修改图书信息

（7）测试根据图书编号删除图书信息的接口。在请求中选择请求方式为 DELETE，输入请求的 URL 地址 http://localhost:8080/chapter06/books，在下方 Params 参数中输入 KEY 和 VALUE。单击 Send 按钮，如图 6-10 所示。

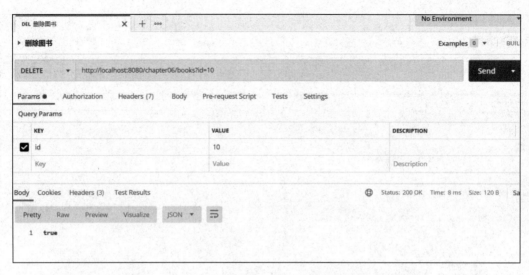

图 6-10　根据图书编号删除图书信息

# 本 章 小 结

本章通过理论和实践相结合的方式，详细介绍了 SSM(Spring、SpringMVC、MyBatis) 的整合开发。通过一个简单的 CRUD 应用，演示了 SSM 整合的全过程，从配置到实现具体功能。SSM 整合为我们提供了一个强大、灵活、可扩展的开发框架，可以应对多样的业务需求。通过本章的学习，读者能够熟练掌握 SSM 整合的基础知识，并在实际项目中灵活应用。

# 课 后 习 题

一、选择题

1. 在 SSM 框架中，Spring 负责(　　)。
　　A. 控制反转和依赖注入　　　　　　B. 处理 HTTP 请求
　　C. 数据库访问　　　　　　　　　　D. 页面渲染
2. 在 SSM 框架中，SpringMVC 负责(　　)。
　　A. 控制反转和依赖注入　　　　　　B. 处理 HTTP 请求
　　C. 数据库访问　　　　　　　　　　D. 页面渲染
3. 在 SSM 框架中，MyBatis 负责(　　)。
　　A. 控制反转和依赖注入　　　　　　B. 处理 HTTP 请求
　　C. 数据库访问　　　　　　　　　　D. 页面渲染

二、填空题

1. 在基于注解的 SSM 整合中，使用 _____ 注解来标记 Spring 配置类，它通常包含被 @ComponentScan 注解扫描的包路径。

2. 在 SpringMVC 中，使用 _____ 注解来标记控制器类，它表示该类是一个 Spring MVC 控制器。

## 三、判断题

1. SSM 整合后，MyBatis 的 SessionFactory 和 Session 对象不需要用户初始化，由 Sping 统一管理。                                （  ）
2. @EnableWebMvc 注解用来在 SSM 整合中启动 SpingMVC 的功能。         （  ）

## 四、简答题

1. 请解释基于注解的 SSM 整合是如何工作的，以及如何使用注解配置 Spring、Spring MVC 和 MyBatis。
2. 在基于注解的 SSM 整合中，如何配置 Spring 扫描注解的包路径，以确保 Spring 能够识别并管理带有注解的组件？

## 五、实操题

创建一个简单的 SSM 项目，内容如下。
(1) 创建一个名为 Student 的实体类，包含 id、name、age、email 字段。
(2) 使用 MyBatis 完成对 Student 表的 CRUD 操作。
(3) 使用 Spring 管理 MyBatis 的 SqlSessionFactory 和数据源。
(4) 使用 SpringMVC 实现 Student 数据的增加、删除、修改和查询功能。
(5) 使用 Postman 工具测试 API。

整合，就是力量

在 SSM 整合开发中，Spring、SpringMVC 和 MyBatis 三者协同工作，实现了更强大的功能。这种整合和协作正是社会主义集体合作理念的体现，在社会生产和发展中，团结和协作同样是成功的关键。

# 第 7 章

# 详解 SpringBoot

**学习目标**

1. 搭建 SpringBoot 项目,学习如何使用 IDEA 创建一个基本的 SpringBoot 项目。
2. 编写一个简单的 SpringBoot 应用程序。
3. 掌握 SpringBoot 的单元测试。

**学习要点**

1. 掌握 SpringBoot 配置的方法。
2. 掌握 SpringBoot 中的 YAML 配置文件的配置方法。
3. 掌握 SpringBoot 的单元测试方法。

本章知识点结构如图 7-1 所示。

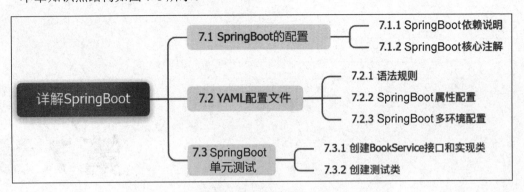

图 7-1　详解 SpringBoot

## 7.1　SpringBoot 的配置

SpringBoot 对于 Java 开发者而言具有显著优势：它不仅大幅简化了 Spring 应用的开发流程，减少了样板代码和配置工作，而且显著提高了开发效率。作为构建微服务架构的理想选择，SpringBoot 在企业级应用开发中得到广泛应用，增强了开发者的职业竞争力。其丰富的生态系统、强大的社区支持、广泛的技术兼容性，以及对各种规模项目的适应能力，都使其成为一个不断更新和改进的现代 Java 开发工具。因此，掌握 SpringBoot，对于追求高效、灵活且先进的应用开发方法的开发者来说，至关重要。

在第 1 章中我们已经搭建了 SpringBoot 的开发环境，下面来回顾一下 SpringBoot 开发环境搭建的过程。

### 7.1.1　SpringBoot 依赖说明

在 IntelliJ IDEA 中创建一个新的 Maven 项目，项目名为 chapter07，在 pom.xml 文件中添加依赖，代码如下：

```xml
<?xml version="1.0" encoding="UTF-8"?>
<project xmlns="http://maven.apache.org/POM/4.0.0"
      xmlns:xsi="http://www.w3.org/2001/XMLSchema-instance"
      xsi:schemaLocation="http://maven.apache.org/POM/4.0.0
        http://maven.apache.org/xsd/maven-4.0.0.xsd">
   <modelVersion>4.0.0</modelVersion>

   <groupId>com.demo</groupId>
   <artifactId>chapter07</artifactId>
   <version>1.0-SNAPSHOT</version>

   <properties>
      <maven.compiler.source>8</maven.compiler.source>
      <maven.compiler.target>8</maven.compiler.target>
      <project.build.sourceEncoding>UTF-8</project.build.sourceEncoding>
   </properties>
```

微课：搭建 SpringBoot 开发环境

```xml
<parent>
    <artifactId>spring-boot-starter-parent</artifactId>
    <groupId>org.springframework.boot</groupId>
    <version>2.6.2</version>
</parent>

<dependencies>
    <dependency>
        <groupId>org.springframework.boot</groupId>
        <artifactId>spring-boot-starter-web</artifactId>
    </dependency>

    <dependency>
        <groupId>org.springframework.boot</groupId>
        <artifactId>spring-boot-starter-test</artifactId>
    </dependency>
</dependencies>
</project>
```

在 Maven 项目中，使用 spring-boot-starter-parent 作为父 POM 的主要作用是提供一个预先配置的依赖和插件管理框架，它简化了版本管理和构建配置，确保了依赖项之间的兼容性，并降低了项目配置的复杂性。这对于标准化项目设置、简化 SpringBoot 应用的构建和维护过程非常有帮助。除了添加父 POM，我们还需要添加两个依赖，它们的作用如下。

（1）spring-boot-starter-web：该依赖提供了构建 Web 应用所需的全部依赖，包括 Spring MVC、Tomcat 作为默认的嵌入式容器以及其他 Web 开发所需的各种工具和库。使用这个依赖，可以快速开发 SpringMVC 应用程序，支持 RESTful 应用程序的开发，并轻松地接入 Spring 框架提供的各种特性，如依赖注入、事务管理等。这是开发基于 SpringBoot 的 Web 应用程序时最常用的依赖之一。

（2）spring-boot-starter-test：该依赖包含了用于测试 SpringBoot 应用程序的各种工具和库。它包括 JUnit、Spring Test、SpringBoot Test、AssertJ、Hamcrest 和其他有用的测试库。通过这个依赖，可以方便地编写和运行单元测试和集成测试，确保代码质量和应用的稳定性。该依赖是 SpringBoot 项目中不可或缺的一部分，对于实现测试驱动开发(TDD)和确保软件质量非常重要。

这两个依赖是 SpringBoot 项目中常见的配置，分别用于简化 Web 层的开发和测试工作。通过这些依赖，SpringBoot 显著降低了项目配置的复杂性，提高了开发效率。

## 7.1.2 SpringBoot 核心注解

在项目的 src/main/java 目录下，创建 com.demo 包，创建 DemoApplication 启动类，代码如下：

```java
package com.demo;

import org.springframework.boot.SpringApplication;
import org.springframework.boot.autoconfigure.SpringBootApplication;
```

```
@SpringBootApplication
public class DemoApplication {

    public static void main(String[] args) {
        SpringApplication.run(DemoApplication.class,args);
    }

}
```

@SpringBootApplication 是一个方便的注解，它在 SpringBoot 应用中经常使用。这个注解本质上是一个复合注解，它将以下三个常用的注解组合在一起。

### 1. @SpringBootConfiguration

标记当前类为配置类，等同于 Spring 的@Configuration 注解。它表明该类使用 Spring 基于 Java 的配置。

### 2. @EnableAutoConfiguration

启用 SpringBoot 的自动配置功能，使 SpringBoot 应用程序根据类路径和依赖关系，自动配置应用程序所需的 Bean，从而减少了开发者的工作量，使得应用程序可以更快速地搭建和运行。例如，如果 spring-webmvc 在类路径上，此注解会标记应用程序为 Web 应用程序并激活关键行为，如设置 DispatcherServlet。

### 3. @ComponentScan

指示 Spring 去扫描其他组件、配置类和服务类等，这样这些组件就可以被自动发现并注册为 Spring 的 Bean。默认情况下，@ComponentScan 会扫描当前包和其子包中的所有类。

将这三个注解组合成一个注解，能够极大地简化 SpringBoot 应用程序的初始引导和配置。通常情况下，开发者在 SpringBoot 应用程序的主类上放置@SpringBootApplication 注解，这样就可以通过运行这个类来启动应用程序。这种做法使得 SpringBoot 应用的创建和运行变得更加简洁和直观。

在项目的 src/main/java 目录下，创建 com.demo.controller 包，创建 DemoController 控制器，代码如下：

```
package com.demo.controller;

import org.springframework.stereotype.Controller;
import org.springframework.web.bind.annotation.RequestMapping;
import org.springframework.web.bind.annotation.ResponseBody;

@Controller
@RequestMapping("/demo")
public class DemoController {

    @RequestMapping("/hello")
    @ResponseBody
    public String hello(){
        System.out.println("hello spring boot");
        return "Hello Spring Boot";
    }

}
```

### 1. 导入的注解

@Controller：这是一个用于声明 SpringMVC 控制器的注解。

@RequestMapping：这个注解用于映射 Web 请求(即 URL 路径)到相应的方法或类上。

@ResponseBody：此注解表明方法的返回值应该直接作为 HTTP 响应正文返回，而不是解释为视图名称。

### 2. 类定义(DemoController)

@Controller 标记这个类为 SpringMVC 的控制器。

@RequestMapping("/demo")指定了这个控制器中所有方法的基础 URL 路径。这意味着此控制器中的所有请求映射都将以/demo 为前缀。

### 3. 方法定义(hello)

@RequestMapping("/hello")注解将 HTTP 请求的/demo/hello 路径映射到这个方法上。

@ResponseBody 意味着这个方法的返回值(在这个例子中是字符串"Hello Spring Boot")将直接作为 HTTP 响应的正文返回给客户端。方法内部打印了"hello spring boot"到控制台，并返回字符串"Hello Spring Boot"。

当访问应用程序的/demo/hello 路径时，这个方法将被调用，并返回文本"Hello Spring Boot"作为响应。这个简单的例子展示了如何在 SpringBoot 应用程序中使用控制器来处理 Web 请求。

打开启动类，单击左侧的 ▷ 按钮，启动项目，在浏览器访问控制器，如图 7-2、图 7-3 所示。

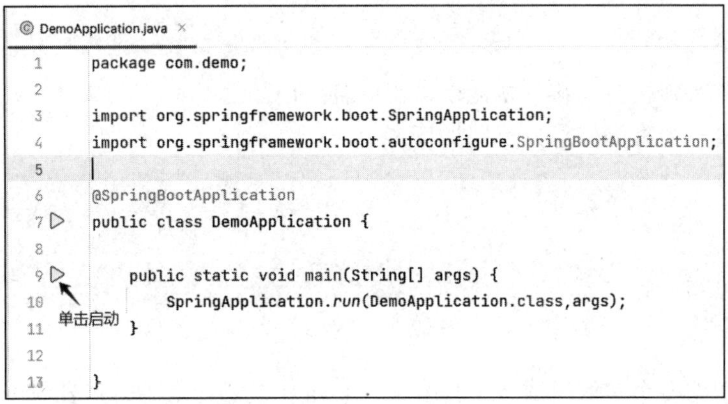

图 7-2 启动项目

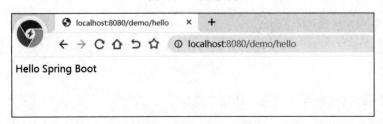

图 7-3 浏览器访问

## 7.2 YAML 配置文件

在 SpringBoot 的应用中，一般使用 YAML 格式的配置文件来对其各项属性进行配置。下面简要介绍 YAML 及 SpringBoot 中的 YAML 配置文件的配置方法。

动画：YAML——优雅的配置管理

### 7.2.1 语法规则

YAML(YAML Ain't Markup Language，YAML 不是标记语言)是一种简洁的数据序列化格式，通常用于配置文件或数据交换。与 JSON 或 XML 相比，YAML 更易读和易写。下面是一些常见的 YAML 语法规则和用法。

微课：YAML 配置文件、多环境配置

**1. 基础语法**

一个 YAML 文件中的内容就是键和值的组合，其中键值对由英文的冒号(:)分割，即 key:value 表示的格式。例如，表示姓名为张三的信息，其结构如下：

```
name: zhangsan
```

**2. 嵌套**

在 YAML 文件中定义信息的时候，有时候需要使这些信息具有层次，例如上例中的姓名属性一般是属于一个人的信息，除了姓名，可能还包含其他信息，此时就需要把这些信息放到人之下，即嵌套。嵌套中的父级信息和子级信息通过两个空格来表示，格式如下：

```
parent:
  child: value
```

注意：缩进必须使用空格，不能使用制表符(tab)。

例如，定义一个人的信息，代码如下：

```
person:
  name: zhangsan
  age: 30
  city: Beijing
```

**3. 列表**

同一个类型的多个实例组成一个列表，与 Java 中的数组类似。列表项以短横线(-)开始。列表示例如下：

```
fruits:
  - Apple
  - Banana
```

**4. 注释**

YAML 中的注释以#字符表示，具体示例如下：

```
# This is a comment
key: value
```

## 7.2.2　SpringBoot 属性配置

SpringBoot 的配置文件名为 application.yml。具体配置步骤和使用如下。

（1）在 resources 目录下创建 application.yml 文件，添加对应配置。name 和 age 属性为自定义属性，在程序中可以通过 key 的名字获取到配置的值。server 下的属性是 SpringBoot 启动自带的服务器使用的配置，context-path 为访问当前 SpringBoot 项目的路径，port 为服务器的监听端口。下面示例中 context-path 为/chapter07，port 为 8989，则访问项目的地址为 http://localhost:8989/chapter07。

```yml
#给控制器的属性赋值
name: zs
age: 20

#配置项目名和端口号
server:
  servlet:
    context-path: /chapter07
  port: 8989
```

（2）在 DemoController 控制器中添加 name、age 属性，使用@Value 注解获取上述配置文件的值，并赋值给对应的属性名，如图 7-4 所示。

```java
import org.springframework.web.bind.annotation.ResponseBody;

@Controller
@RequestMapping("/demo")
public class DemoController {

    @Value("${name}")
    private String name;

    @Value("${age}")
    private int age;

    @RequestMapping("/test")
    @ResponseBody
    public String test(){
        return "name = " + name + ", age = " + age;
    }
}
```

图 7-4　属性赋值

（3）重新启动项目，并且访问控制器，注意端口号和项目名称，返回结果中显示了配置文件中自定义的属性 name 和 age 的值，如图 7-5 所示。

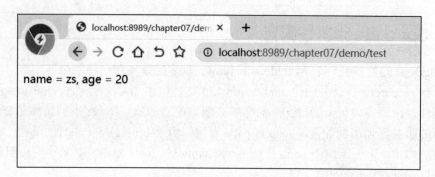

图 7-5  浏览器访问测试

### 7.2.3  SpringBoot 多环境配置

在实际开发中往往存在多个程序运行环境，如开发环境、测试环境、生产环境等。配置文件的参数如果经常改来改去，非常麻烦且容易出错，此时使用多环境配置就显得非常有必要了。在 application.yml 文件中配置多个环境，并且通过 spring.profile.active 激活 dev 环境，如图 7-6 所示。

```yaml
name: zs
age: 20


#server:
#  servlet:
#    context-path: /chapter07
#  port: 8989

spring:
  profiles:
    active: dev

---
spring:
  config:
    activate:
      on-profile: pro
server:
  port: 8081

---
spring:
  config:
    activate:
      on-profile: dev
server:
  port: 8082
```

图 7-6  多环境配置

启动项目，可以看到控制台显示的端口号为 8082，如图 7-7 所示。

图 7-7　控制台显示端口号

打开浏览器访问项目地址，结果如图 7-8 所示。

图 7-8　浏览器访问项目地址

## 7.3　SpringBoot 单元测试

微课：单元测试

SpringBoot 项目开发过程中，可以针对业务逻辑层单独测试。详细步骤如下。

### 7.3.1　创建 BookService 接口和实现类

在 src/main/java 目录下创建 com.demo.service 包，在该包下创建 BookService 接口，代码如下：

```
package com.demo.service;
public interface BookService {
    void save();
}
```

在 src/main/java 目录下创建 com.demo.service.impl 包，在该包下创建 BookServiceImpl 实现类，代码如下：

```
package com.demo.service.impl;

import com.demo.service.BookService;
import org.springframework.stereotype.Service;

@Service
public class BookServiceImpl implements BookService {
    @Override
    public void save() {
        System.out.println("业务层：保存图书");
```

    }
}

## 7.3.2 创建测试类

在 src/test/java 目录下创建 com.demo.test 包，在该包下创建 BookServiceTest 测试类。代码如下：

```java
package com.demo.test;

import com.demo.service.BookService;
import org.junit.jupiter.api.Test;
import org.springframework.beans.factory.annotation.Autowired;
import org.springframework.boot.test.context.SpringBootTest;

@SpringBootTest
public class BookServiceTest {

    @Autowired
    private BookService bookService;

    @Test
    public void testSave() {
        bookService.save();
    }

}
```

单击左侧按钮，运行对应的方法，观察控制台输出结果，如图 7-9、图 7-10 所示。

图 7-9　单元测试

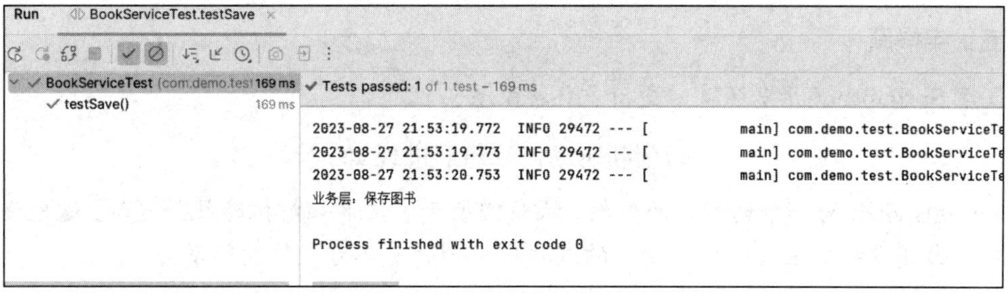

图 7-10　运行结果

# 本 章 小 结

在本章中，我们学习了 SpringBoot 的基础知识，了解了 SpringBoot 的优势，以及如何使用 IDEA 快速创建一个 SpringBoot 项目，并学习了 YAML 配置文件的应用以及单元测试。

# 课 后 习 题

**一、选择题**

1. SpringBoot 是用于(　　)开发的。
   A. 移动应用程序　　B. Web 应用程序　　C. 桌面应用程序　　D. 游戏应用程序
2. 在 SpringBoot 中，通常使用(　　)文件来配置应用程序的属性。
   A. application.properties　　　　　　B. application.yml
   C. application.xml　　　　　　　　　D. application.config
3. 在 SpringBoot 中，用于编写和运行单元测试的框架是(　　)。
   A. JUnit　　　　　B. TestNG　　　　　C. Mockito　　　　　D. Cucumber

**二、填空题**

1. 要搭建 SpringBoot 环境，需要在项目中添加 SpringBoot 的_____依赖。
2. 在 SpringBoot 中，应用程序的配置通常存储在_____或_____文件中。

**三、判断题**

1. SpringBoot 仅适用于 Web 应用程序的开发，不支持其他类型的应用程序。　　(　　)
2. SpringBoot 应用程序的配置文件可以使用 .properties 文件或 .yml 文件进行配置。
   　　　　　　　　　　　　　　　　　　　　　　　　　　　　　　　　　　(　　)

**四、简答题**

1. 解释什么是 SpringBoot，以及它的主要目标和优点。
2. 什么是单元测试，为什么在开发 SpringBoot 应用程序时编写单元测试是很重要的？示例说明如何编写 SpringBoot 应用程序的单元测试。

五、实操题

搭建 SpringBoot 开发环境,支持多环境配置。

创新驱动,梦想照进现实

SpringBoot 作为一个创新性的框架,给我们展示了技术如何推动社会进步。这也提醒我们,在全面建设社会主义现代化国家的过程中,创新是不可或缺的驱动力。

# 第 8 章
# SpringBoot 集成 MyBatis

**学习目标**

1. 掌握 SpringBoot 集成 MyBatis 的步骤。
2. 学习如何在 SpringBoot 项目中配置和集成 MyBatis，以便使用 MyBatis 进行数据访问。
3. 学习配置数据源：了解如何配置数据源(DataSource)，并将其与 MyBatis 集成，以便连接数据库。
4. 编写 MyBatis 映射文件：学习如何编写 MyBatis 映射文件(Mapper)，定义 SQL 查询和操作。

**学习要点**

1. 掌握 SpringBoot 整合 MyBatis 的开发流程。
2. 掌握 SpringBoot 和 MyBatis 的增删改查操作。

本章知识点结构如图 8-1 所示。

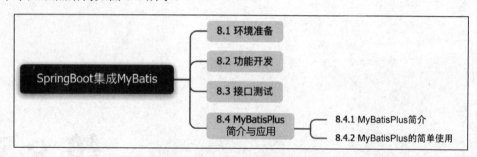

图 8-1 SpringBoot 集成 MyBatis

SpringBoot 与 MyBatis 的整合提供了一种高效且简便的方式来开发基于 Java 的数据库应用程序。SpringBoot 是一个用于简化 Spring 应用开发的框架，而 MyBatis 是一个流行的 SQL 映射框架，它提供了一种相对简单的方法来访问数据库。当这两个框架结合使用时，具有以下优势。

### 1．简化配置

SpringBoot 的自动配置功能减少了需要编写的配置代码。与 MyBatis 整合时，大多数配置都是自动完成的，包括数据源和事务管理。

动画：SpringBoot 与 MyBatis 完美融合：构建高效数据访问

### 2．强大的数据访问

MyBatis 允许开发者直接编写 SQL，从而可以精确地控制数据访问逻辑。这对于那些需要编写复杂查询或优化数据库性能的场景非常有用。

### 3．映射灵活性

MyBatis 提供了强大的映射能力，允许将数据库结果映射到 Java 对象上。这降低了数据处理的复杂性。

### 4．集成支持

SpringBoot 和 MyBatis 易于与其他 Spring 生态系统组件(如 Spring Security、Spring Data 等)集成，使得构建完整的应用程序更加方便。

### 5．社区和资源

由于 Spring 和 MyBatis 都拥有活跃的社区，开发者可以轻松地找到相关的教程、最佳实践和问题解决方案。

下面将通过一个简单商品信息管理项目，来演示 SpringBoot 和 MyBatis 的整合过程。

## 8.1 环境准备

(1) 在 IntelliJ IDEA 中创建一个新的 Maven 项目，项目名为 chapter08，在 pom.xml 文件中添加依赖，代码如下：

微课：搭建 SpringBoot 和 MyBatis 基础环境

```xml
<?xml version="1.0" encoding="UTF-8"?>
<project xmlns="http://maven.apache.org/POM/4.0.0"
```

```xml
    xmlns:xsi="http://www.w3.org/2001/XMLSchema-instance"
    xsi:schemaLocation="http://maven.apache.org/POM/4.0.0
    http://maven.apache.org/xsd/maven-4.0.0.xsd">
<modelVersion>4.0.0</modelVersion>

<groupId>com.demo</groupId>
<artifactId>chapter08</artifactId>
<version>1.0-SNAPSHOT</version>

<properties>
    <maven.compiler.source>8</maven.compiler.source>
    <maven.compiler.target>8</maven.compiler.target>
    <project.build.sourceEncoding>UTF-8</project.build.sourceEncoding>
</properties>

<parent>
    <artifactId>spring-boot-starter-parent</artifactId>
    <groupId>org.springframework.boot</groupId>
    <version>2.6.2</version>
</parent>

<dependencies>
    <!--SpringBoot集成SpringMVC-->
    <dependency>
        <groupId>org.springframework.boot</groupId>
        <artifactId>spring-boot-starter-web</artifactId>
    </dependency>
    <!--SpringBoot测试-->
    <dependency>
        <groupId>org.springframework.boot</groupId>
        <artifactId>spring-boot-starter-test</artifactId>
    </dependency>

    <!--SpringBoot集成MyBatis-->
    <dependency>
        <groupId>org.mybatis.spring.boot</groupId>
        <artifactId>mybatis-spring-boot-starter</artifactId>
        <version>2.2.0</version>
    </dependency>

    <!-- 数据库连接池 -->
    <dependency>
        <groupId>com.alibaba</groupId>
        <artifactId>druid</artifactId>
        <version>1.2.6</version>
    </dependency>

    <!-- Lombok自动生成方法功能 -->
    <dependency>
        <groupId>org.projectlombok</groupId>
        <artifactId>lombok</artifactId>
    </dependency>
```

```xml
        <!-- MySQL 驱动 -->
        <dependency>
            <groupId>mysql</groupId>
            <artifactId>mysql-connector-java</artifactId>
            <version>5.1.43</version>
        </dependency>
    </dependencies>

</project>
```

(2) 创建数据库和数据表，添加测试数据。下面的 SQL 创建了数据库 springboot_demo，并在数据库中创建商品信息表 product，包含商品标识 id、名称 name、价格 price 和描述 description。

```sql
CREATE DATABASE springboot_demo CHARACTER SET utf8;

USE springboot_demo;

CREATE TABLE product(
    id INT PRIMARY KEY AUTO_INCREMENT,
    name VARCHAR(255),
    price DECIMAL(10, 2),
    description VARCHAR(255)
);
```

数据表创建之后，向表中插入一些测试数据，具体 SQL 语句如下：

```sql
-- 向 product 表中插入商品数据
INSERT INTO product (name, price, description) VALUES
('笔记本电脑', 9999.99, '高性能笔记本，配备 16GB RAM'),
('智能手机', 6999.99, '128GB 存储的 Android 智能手机'),
('咖啡机', 499.99, '自动滴漏式咖啡机'),
('办公椅', 1999.99, '符合人体工学的办公椅'),
('背包', 899.99, '带 USB 充电口的笔记本电脑背包'),
('电视机', 5999.99, '55 英寸 4K UHD 智能电视'),
('智能手表', 2499.99, '健身和健康追踪器');
```

(3) 在 resources 目录下创建 application.yml 文件，除了第 7 章中介绍过的 SpringBoot 服务器的配置之外，如下配置文件中还添加了 MySQL 数据库连接的信息 datasource 部分和 MyBatis 相关的配置 MyBatis 部分，详细配置如下：

```yaml
#项目访问名称
server:
  servlet:
context-path: /chapter08

spring:
  # 数据库连接池的配置
  datasource:
    type: com.alibaba.druid.pool.DruidDataSource
    driver-class-name: com.mysql.jdbc.Driver
```

```yaml
    url: jdbc:mysql://localhost:3306/springboot_demo?useSSL=false&characterEncoding=utf8
    username: root
    password: 123456

# MyBatis 的配置
mybatis:
  type-aliases-package: com.demo.pojo
  mapper-locations:
    - mappers/*Mapper.xml
  configuration:
    map-underscore-to-camel-case: true
    log-impl: org.apache.ibatis.logging.stdout.StdOutImpl
```

(4) 在项目的 src/main/java 目录下，创建 com.demo 包，创建 SpringBoot 启动类 DemoApplication，代码如下：

```java
package com.demo;

import org.mybatis.spring.annotation.MapperScan;
import org.springframework.boot.SpringApplication;
import org.springframework.boot.autoconfigure.SpringBootApplication;

@SpringBootApplication
@MapperScan("com.demo.mapper")
public class DemoApplication {

    public static void main(String[] args) {
        SpringApplication.run(DemoApplication.class,args);
    }

}
```

## 8.2 功能开发

整个项目结构配置搭建好之后，可以进行项目的数据层、业务层和控制器层的开发，具体步骤如下。

微课：SpringBoot+MyBatis 的功能开发

(1) 在项目的 src/main/java 目录下，创建 com.demo.pojo 包，创建 Product 实体类，代码如下：

```java
package com.demo.pojo;

import lombok.Data;

import java.math.BigDecimal;

@Data
public class Product {
```

```
    private Long id;
    private String name;
    private BigDecimal price;
    private String description;
}
```

(2) 在项目的 src/main/java 目录下，创建 com.demo.mapper 包，创建 ProductMapper 接口，代码如下：

```
package com.demo.mapper;

import com.demo.pojo.Product;
import org.apache.ibatis.annotations.Param;

import java.util.List;

public interface ProductMapper {
    List<Product> getAllProducts();
    Product getProductById(@Param("id") Long id);
    void createProduct(Product product);
    void updateProduct(Product product);
    void deleteProduct(@Param("id") Long id);
}
```

(3) 在项目的 src/main/java 目录下，创建 com.demo.service 包，创建 ProductService 接口，代码如下：

```
package com.demo.service;

import com.demo.pojo.Product;

import java.util.List;

public interface ProductService {
    List<Product> getAllProducts();
    Product getProductById(Long id);
    boolean createProduct(Product product);
    boolean updateProduct(Product product);
    boolean deleteProduct(Long id);
}
```

(4) 在项目的 src/main/java 目录下，创建 com.demo.service.impl 包，创建 ProductServiceImpl 接口，代码如下：

```
package com.demo.service.impl;

import com.demo.mapper.ProductMapper;
import com.demo.pojo.Product;
import com.demo.service.ProductService;
import org.springframework.beans.factory.annotation.Autowired;
```

```java
import org.springframework.stereotype.Service;
import java.util.List;

@Service
public class ProductServiceImpl implements ProductService {

    @Autowired
    private ProductMapper productMapper;

    @Override
    public List<Product> getAllProducts() {
        return productMapper.getAllProducts();
    }

    @Override
    public Product getProductById(Long id) {
        return productMapper.getProductById(id);
    }

    @Override
    public boolean createProduct(Product product) {
        productMapper.createProduct(product);
        return true;
    }

    @Override
    public boolean updateProduct(Product product) {
        productMapper.updateProduct(product);
        return true;
    }

    @Override
    public boolean deleteProduct(Long id) {
        productMapper.deleteProduct(id);
        return true;
    }

}
```

(5) 在项目的 src/main/java 目录下，创建 com.demo.controller 包，创建 ProductController 类，代码如下：

```java
package com.demo.controller;

import com.demo.pojo.Product;
import com.demo.service.ProductService;
import org.springframework.beans.factory.annotation.Autowired;
import org.springframework.stereotype.Controller;
import org.springframework.web.bind.annotation.RequestMapping;
import org.springframework.web.bind.annotation.RequestMethod;
```

```java
import org.springframework.web.bind.annotation.ResponseBody;

import java.util.List;

@Controller
@RequestMapping("/products")
public class ProductController{

    @Autowired
    private ProductService productService;

    @RequestMapping(method = RequestMethod.POST)
    @ResponseBody
    public boolean save(Product product) {
        return productService.createProduct(product);
    }

    @RequestMapping(method = RequestMethod.PUT)
    @ResponseBody
    public boolean update(Product product) {
        return productService.updateProduct(product);
    }

    @RequestMapping(method = RequestMethod.DELETE)
    @ResponseBody
    public boolean delete(Long id) {
        return productService.deleteProduct(id);
    }

    @RequestMapping(method = RequestMethod.GET,value = "getById")
    @ResponseBody
    public Product getById(Long id) {
        return productService.getProductById(id);
    }

    @RequestMapping(method = RequestMethod.GET)
    @ResponseBody
    public List<Product> getAll() {
        return productService.getAllProducts();
    }
}
```

## 8.3 接口测试

代码开发完成之后，启动项目，使用 Postman 对接口进行测试，这里只对查询商品接口做测试，其他接口的测试读者可以自己进行，查询接口结果如图 8-2 所示。

图 8-2　Postman 接口测试

## 8.4　MyBatisPlus 简介与应用

微课：MyBatisPlus 开发

### 8.4.1　MyBatisPlus 简介

MyBatisPlus 是一个 MyBatis 的增强工具，在 MyBatis 的基础上只做增强不做改变，为简化开发和提高效率而生。它内置了多种常用的功能，如分页插件、通用 CRUD 操作、条件构造器等，让 MyBatis 的操作更加便捷。

### 8.4.2　MyBatisPlus 的简单使用

下面通过一个简单示例，演示 MyBatisPlus 在 SpringBoot 中的使用。

(1) 在 pom.xml 文件中添加 MyBatisPlus 的 Maven 依赖，代码如下：

```xml
<dependency>
    <groupId>com.baomidou</groupId>
    <artifactId>mybatis-plus-boot-starter</artifactId>
    <version>3.5.0</version>
</dependency>
```

(2) 创建一个用户表 user，表中包含三个字段，即用户标号 id、姓名 name 和年龄 age，代码如下：

```sql
CREATE TABLE `user` (
 `id` INT(11) NOT NULL AUTO_INCREMENT,
 `name` VARCHAR(50) NOT NULL,
 `age` INT(11) NOT NULL,
```

```
    PRIMARY KEY (`id`)
);
```

(3) 创建一个与 user 表对应的 User 实体类，并添加 MyBatisPlus 的注解，代码如下：

```java
import com.baomidou.mybatisplus.annotation.IdType;
import com.baomidou.mybatisplus.annotation.TableId;
import com.baomidou.mybatisplus.annotation.TableName;
import lombok.Data;

@TableName("user")
@Data
public class User {
    @TableId(type = IdType.AUTO)
    private Integer id;
    private String name;
    private Integer age;
}
```

(4) 创建一个 UserMapper 接口，并继承 BaseMapper<User>，代码如下：

```java
import com.baomidou.mybatisplus.core.mapper.BaseMapper;
import org.apache.ibatis.annotations.Mapper;

@Mapper
public interface UserMapper extends BaseMapper<User> {
}
```

(5) 创建一个 UserService 接口，代码如下：

```java
package com.demo.service;

import com.baomidou.mybatisplus.extension.plugins.pagination.Page;
import com.demo.pojo.User;

import java.util.List;

public interface UserService {

    public void createUser(User user);

    public User readUser(int id);

    public void updateUser(User user);

    public void deleteUser(int id);

    public Page<User> listUsersByPage(int pageNo, int pageSize);

    public List<User> listUsersByName(String name);

}
```

(6) 创建 UserServiceImpl 实现类，代码如下：

```java
package com.demo.service.impl;

import com.baomidou.mybatisplus.extension.plugins.pagination.Page;
import com.demo.mapper.UserMapper;
import com.demo.pojo.User;
import com.demo.service.UserService;
import com.baomidou.mybatisplus.core.conditions.query.QueryWrapper;
import org.springframework.beans.factory.annotation.Autowired;
import org.springframework.stereotype.Service;

import java.util.List;

@Service
public class UserServiceImpl implements UserService {

    @Autowired
    private UserMapper userMapper;

    @Override
    public void createUser(User user) {
        userMapper.insert(user);
    }

    @Override
    public User readUser(int id) {
        return userMapper.selectById(id);
    }

    @Override
    public void updateUser(User user) {
        userMapper.updateById(user);
    }

    @Override
    public void deleteUser(int id) {
        userMapper.deleteById(id);
    }

    @Override
    public Page<User> listUsersByPage(int pageNo, int pageSize) {
        //对分页进行封装
        Page<User> page = new Page<>(pageNo, pageSize);
        return userMapper.selectPage(page, null);
    }

    public List<User> listUsersByName(String name) {
        //存储查询条件
        QueryWrapper<User> queryWrapper = new QueryWrapper<>();
        queryWrapper.eq("name", name);
```

```
        return userMapper.selectList(queryWrapper);
    }
}
```

(7) 创建 UserService Test 测试类，代码如下：

```
package com.test;

import com.baomidou.mybatisplus.extension.plugins.pagination.Page;
import com.demo.DemoApplication;
import com.demo.pojo.User;
import com.demo.service.UserService;
import org.junit.jupiter.api.Test;
import org.springframework.beans.factory.annotation.Autowired;
import org.springframework.boot.test.context.SpringBootTest;

import java.util.List;


@SpringBootTest(classes = DemoApplication.class)
public class UserServiceTest {

    @Autowired
    private UserService userService;

    @Test
    public void testCreateUser() {
        User user = new User();
        user.setName("Alice");
        user.setAge(25);
        userService.createUser(user);
    }

    @Test
    public void testReadUser() {
        User user = userService.readUser(1);
        System.out.println(user);
    }

    @Test
    public void testUpdateUser() {
        User user = new User();
        user.setId(1);
        user.setName("Bob");
        userService.updateUser(user);
    }

    @Test
    public void testDeleteUser() {
        userService.deleteUser(1);
```

```java
    }

    @Test
    public void testListUsersByPage() {
        Page<User> page = userService.listUsersByPage(1, 5);
        List<User> list = page.getRecords();
        for (User user : list) {
            System.out.println(user);
        }
    }

    @Test
    public void testListUsersByName() {
        List<User> users = userService.listUsersByName("Alice");
        for (User user : users) {
            System.out.println(user);
        }
    }
}
```

(8) 单击左侧按钮，运行对应的方法，观察控制台输出结果，如图 8-3～图 8-5 所示。

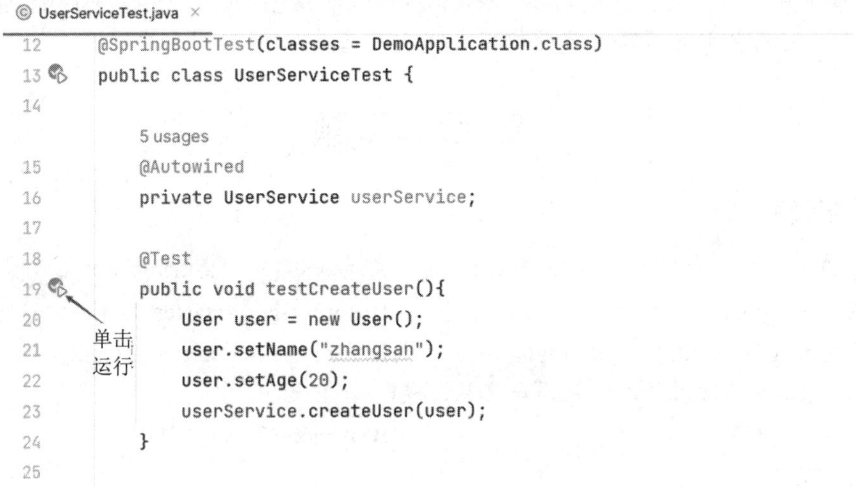

```
@SpringBootTest(classes = DemoApplication.class)
public class UserServiceTest {

    5 usages
    @Autowired
    private UserService userService;

    @Test
    public void testCreateUser(){
        User user = new User();
        user.setName("zhangsan");
        user.setAge(20);
        userService.createUser(user);
    }
```

单击
运行

图 8-3　运行创建用户方法

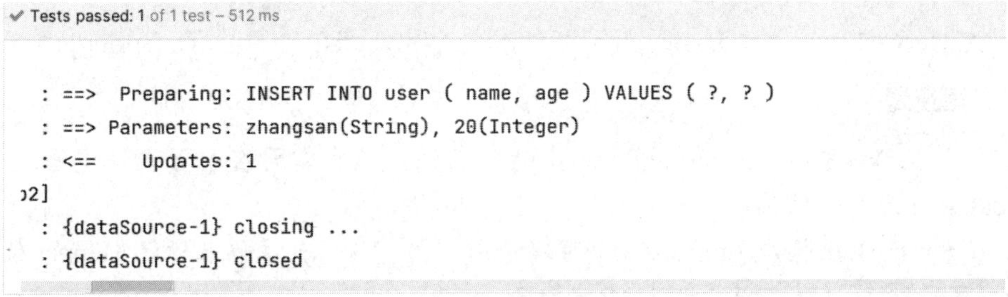

```
✓ Tests passed: 1 of 1 test – 512 ms

  : ==>  Preparing: INSERT INTO user ( name, age ) VALUES ( ?, ? )
  : ==> Parameters: zhangsan(String), 20(Integer)
  : <==    Updates: 1
)2]
  : {dataSource-1} closing ...
  : {dataSource-1} closed
```

图 8-4　控制台输出结果

图 8-5　数据库用户表查询结果

UserServiceTest 类中其他方法的测试方式和上述方式相同，这里不再一一罗列。

## 本 章 小 结

在本章中，我们深入学习了如何在 SpringBoot 中集成 MyBatis，以便更加灵活地进行数据库操作，了解了 MyBatis 是一个用于简化持久层操作的框架，可以通过 XML 配置文件来定义 SQL 映射和数据访问。通过将 MyBatis 与 SpringBoot 结合，我们可以充分利用 SpringBoot 的便捷特性来简化项目配置和开发流程。

通过本章的学习，读者已经具备了在 SpringBoot 中使用 MyBatis 进行数据库操作的基本能力，为后续的开发和项目实践打下了坚实的基础。

## 课 后 习 题

一、选择题

1. 在 SpringBoot 中，要集成 MyBatis，通常需要添加(　　)SpringBoot 启动器依赖。

　　A. spring-boot-starter-web　　　　　　B. spring-boot-starter-data-jpa

　　C. spring-boot-starter-data-mybatis　　 D. spring-boot-starter-thymeleaf

2. SpringBoot 中使用的默认数据库连接池是(　　)。

　　A. HikariCP　　　　　　　　　　　　B. Apache Tomcat JDBC

　　C. C3P0　　　　　　　　　　　　　　D. DBCP

3. 在 SpringBoot 应用程序的配置文件中，用于配置 MyBatis 的属性通常位于(　　)部分。

　　A. [server]　　　B. [logging]　　　C. [mybatis]　　　D. [spring]

二、填空题

1. 要配置 MyBatis 在 SpringBoot 中的数据源，通常需要在配置文件中添加一个以 spring.datasource 开头的属性，例如_____ 和 _____。

2. 在 SpringBoot 集成 MyBatis 时，可以使用 _____ 注解来标记 Mapper 接口，以便 SpringBoot 能够自动扫描并创建 Mapper Bean。

### 三、判断题

1. SpringBoot 中集成 MyBatis 时，需要手动配置数据库连接池，而不使用默认的 HikariCP。（　　）

2. 在 SpringBoot 中集成 MyBatis 时，需要手动创建 Mapper 接口的实例，而不是让 SpringBoot 自动扫描和管理它们。（　　）

### 四、简答题

1. 请解释如何在 SpringBoot 应用程序中集成 MyBatis，并提供关键配置的示例。

2. 请描述在 SpringBoot 集成 MyBatis 时，如何处理事务管理以确保数据的一致性。

### 五、实操题

设计一个简单的学生管理系统，练习内容如下。

(1) 创建学生实体类(包含学号、姓名、年龄等属性)，并使用 MyBatis 注解方式配置 Mapper 接口和 SQL 查询。

(2) 实现学生信息的增删改查功能，涵盖 MyBatis 基本的数据操作。

## 技术创新与社会责任

在这个数字化时代，技术的迅猛发展正在改变着我们的生活，而 SpringBoot 这样的创新性框架成为了技术驱动社会进步的典范。然而，除了技术本身，我们还应该认识到在全面建设社会主义现代化国家的过程中，创新是不可或缺的驱动力。

创新不仅仅是技术领域的事情，它贯穿于社会的方方面面。从经济结构的转型升级到教育体系的改革创新，从科学研究的突破到社会治理的现代化，都离不开创新的推动力。

然而，创新也伴随着伦理和法律的挑战。在追求技术进步的同时，我们需要确保创新不侵犯到个人隐私等权利。这就需要技术开发者与法律顾问、伦理专家紧密合作，建立合适的监管和规范，以确保社会的可持续发展。

因此，作为现代社会的一部分，我们不仅要关注技术创新的成果，还要思考如何将创新的力量引导到社会责任感和道德规范上。只有在技术驱动和社会责任之间找到平衡，我们才能够实现梦想，让创新成果走进现实，为社会的进步和现代化之路注入持久的力量。这正是我们全面建设社会主义现代化国家的使命和责任。

# 第 9 章
# 过滤器、拦截器、文件上传和缓存

**学习目标**

1. 了解过滤器和拦截器的概念及其在 SpringBoot 中的应用。
2. 学习如何创建和配置自定义过滤器。
3. 学习如何创建和配置自定义拦截器。
4. 实现文件上传功能。

**学习要点**

1. 过滤器的作用和原理。
2. 拦截器的概念和用途。
3. SpringBoot 中的文件上传。

本章知识点结构如图 9-1 所示。

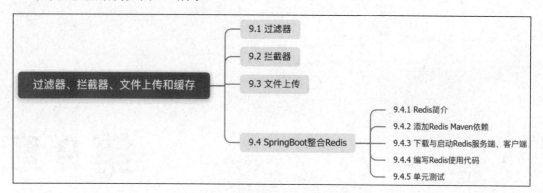

图 9-1　过滤器、拦截器、文件上传和缓存

## 9.1　过　滤　器

动画：探索 SpringBoot 的过滤器——构建更安全、更高效的应用

过滤器(Filter)是 Java Web 应用程序中的一个关键组件，用于在 HTTP 请求到达 Servlet 之前或 HTTP 响应返回给客户端之前，对请求和响应进行拦截、处理和修改。过滤器允许开发人员在 Web 应用程序的处理流程中插入自定义逻辑，以执行一系列任务，如数据处理、请求验证、日志记录等。

下面演示一个简单过滤器的使用。

微课：过滤器

(1) 在 IntelliJ IDEA 中创建一个新的 Maven 项目，项目名为 chapter09，在 pom.xml 文件中添加依赖，代码如下：

```xml
<?xml version="1.0" encoding="UTF-8"?>
<project xmlns="http://maven.apache.org/POM/4.0.0"
    xmlns:xsi="http://www.w3.org/2001/XMLSchema-instance"
    xsi:schemaLocation="http://maven.apache.org/POM/4.0.0
    http://maven.apache.org/xsd/maven-4.0.0.xsd">
    <modelVersion>4.0.0</modelVersion>

    <groupId>com.demo</groupId>
    <artifactId>chapter09</artifactId>
    <version>1.0-SNAPSHOT</version>

    <properties>
        <maven.compiler.source>8</maven.compiler.source>
        <maven.compiler.target>8</maven.compiler.target>
        <project.build.sourceEncoding>UTF-8</project.build.sourceEncoding>
    </properties>

    <parent>
        <artifactId>spring-boot-starter-parent</artifactId>
        <groupId>org.springframework.boot</groupId>
        <version>2.6.2</version>
```

```xml
    </parent>

    <dependencies>
        <dependency>
            <groupId>org.springframework.boot</groupId>
            <artifactId>spring-boot-starter-web</artifactId>
        </dependency>
        <!-- 过滤器所需的依赖 -->
        <dependency>
            <groupId>javax.servlet</groupId>
            <artifactId>javax.servlet-api</artifactId>
            <version>3.1.0</version>
            <scope>provided</scope>
        </dependency>
    </dependencies>
</project>
```

（2）在项目的 src/main/java 目录下，创建 com.demo 包，创建 DemoApplication 启动类，代码如下：

```java
package com.demo;

import org.springframework.boot.SpringApplication;
import org.springframework.boot.autoconfigure.SpringBootApplication;
import org.springframework.boot.web.servlet.ServletComponentScan;

@SpringBootApplication
@ServletComponentScan
public class DemoApplication {

    public static void main(String[] args) {
        SpringApplication.run(DemoApplication.class,args);
    }

}
```

（3）在项目的 src/main/java 目录下，创建 com.demo.controller 包，创建 UserController 控制器类，代码如下：

```java
package com.demo.controller;

import org.springframework.web.bind.annotation.GetMapping;
import org.springframework.web.bind.annotation.RequestMapping;
import org.springframework.web.bind.annotation.RestController;

@RestController
@RequestMapping("user")
public class UserController {

    @GetMapping("getUserInfo")
    public String getUserInfo(){
        System.out.println("getUserInfo");
```

```
        return "ok";
    }

    @GetMapping("login")
    public String login(){
        System.out.println("login");
        return "ok";
    }

}
```

（4）在项目的 src/main/java 目录下，创建 com.demo.filter 包，创建 TokenFilter 过滤器类，TokenFilter 要覆写 Filter 接口的 doFilter 方法实现过滤功能，代码如下：

```
package com.demo.filter;

import com.fasterxml.jackson.databind.ObjectMapper;

import javax.servlet.*;
import javax.servlet.annotation.WebFilter;
import javax.servlet.http.HttpServletRequest;
import java.io.IOException;
import java.util.HashMap;
import java.util.Map;

//配置 Filter 要过滤的 URL，/user/*表示所有以此为开头的 URL 都要经过本过滤器过滤
@WebFilter(urlPatterns = "/user/*", filterName = "tokenFilter")
public class TokenFilter  implements Filter {

    @Override
    public void init(FilterConfig filterConfig) throws ServletException {
        System.out.println(" init ");
    }

    @Override
    public void doFilter(ServletRequest servletRequest,
      ServletResponse servletResponse, FilterChain filterChain)
            throws IOException, ServletException {
        System.out.println(" doFilter ");
        HttpServletRequest request = (HttpServletRequest) servletRequest;
        //从请求头中获取 token，如果没有 token 表示验证不通过，过滤器返回错误
        String token = request.getHeader("token");
        if(token!=null){
            filterChain.doFilter(servletRequest, servletResponse);
        }else{
            servletResponse.setCharacterEncoding("UTF-8");
            servletResponse.setContentType
                ("application/json; charset=utf-8");
            Map<String,String> resultMap= new HashMap<>();
            resultMap.put("msg","错误");
            ObjectMapper mapper = new ObjectMapper();
            mapper.writeValue(servletResponse.getWriter(),resultMap);
```

```
        }
    }

    @Override
    public void destroy() {
        System.out.println(" destroy ");
    }
}
```

(5) 在 Postman 中创建一个 GET 请求，请求地址是 http://localhost:8080/user/getUserInfo。响应一个 JSON 格式的字符串，提示信息为错误，如图 9-2 所示。

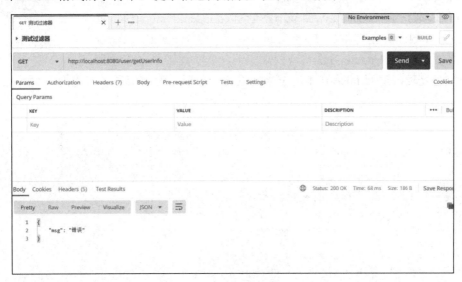

图 9-2　过滤器测试未通过

(6) 在 Headers 选项卡中添加一对请求头参数。KEY 是 token，VALUE 任意输入。这一次请求结果为 ok，说明通过了过滤器，请求到了 Controller 控制器，如图 9-3 所示。

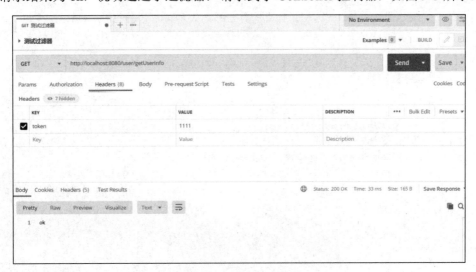

图 9-3　过滤器测试通过

## 9.2 拦截器

拦截器(Interceptor)是一种用于拦截、处理和增强 HTTP 请求和响应的组件,主要用于 Web 应用程序中。拦截器允许在请求到达控制器之前或响应返回给客户端之前执行自定义逻辑。拦截器通常与 Web 框架(如 SpringMVC)相关联,用于实现横切关注点(cross-cutting concerns)的处理。

动画:SpringBoot 拦截器——构建更加稳健的应用

下面演示一个简单拦截器的使用。

(1) 在项目的 src/main/java 目录下,创建 com.demo.interceptor 包,创建 LoginInterceptor 拦截器类,LoginInterceptor 类实现 HandlerInterceptor 接口并且要覆写 preHandle 方法实现权限验证,方法返回 true 表示请求可以继续处理,返回 false 表示请求未通过此拦截器,返回错误,完整代码如下:

微课:拦截器

```java
package com.demo.interceptor;

import com.fasterxml.jackson.databind.ObjectMapper;
import org.springframework.web.servlet.HandlerInterceptor;
import org.springframework.web.servlet.ModelAndView;

import javax.servlet.http.HttpServletRequest;
import javax.servlet.http.HttpServletResponse;
import java.util.HashMap;
import java.util.Map;

public class LoginInterceptor implements HandlerInterceptor {

    @Override
    public boolean preHandle(HttpServletRequest request, HttpServletResponse
        response, Object handler) throws Exception {
        System.out.println(" preHandle ");
        //从请求头中获取token,如果没有token表示验证不通过,过滤器返回错误
        String token = request.getHeader("token");
        if(token!=null){
            return true;
        }else{
            response.setCharacterEncoding("UTF-8");
            response.setContentType("application/json; charset=utf-8");
            Map<String,String> resultMap = new HashMap<>();
            resultMap.put("msg","错误");
            ObjectMapper mapper = new ObjectMapper();
            mapper.writeValue(response.getWriter(),resultMap);
        }
        return false;
    }

    @Override
    public void postHandle(HttpServletRequest request, HttpServletResponse
        response, Object handler, ModelAndView modelAndView)
        throws Exception {
```

```java
        System.out.println(" postHandle ");
    }

    @Override
    public void afterCompletion(HttpServletRequest request, HttpServletResponse
        response, Object handler, Exception ex) throws Exception {
        System.out.println(" afterCompletion ");
    }
}
```

(2) 在项目的 src/main/java 目录下，创建 com.demo.config 包，创建 MyMvcConfig 拦截器配置类，MyMvcConfig 实现 SpringMVC 框架的 WebMvcConfigurer 接口用来注册拦截器，并且调用 addPathPatterns 配置要拦截的 URL 和调用 excludePathPatterns 配置不拦截的 URL，完整代码如下：

```java
package com.demo.config;

import com.demo.interceptor.LoginInterceptor;
import org.springframework.context.annotation.Bean;
import org.springframework.context.annotation.Configuration;
import org.springframework.web.servlet.config.annotation.InterceptorRegistry;
import org.springframework.web.servlet.config.annotation.WebMvcConfigurer;

@Configuration
public class MyMvcConfig implements WebMvcConfigurer {
    //自动注入LoginInterceptor 类的对象
    @Bean
    public LoginInterceptor getInterceptor(){
        return new LoginInterceptor();
    }

    @Override
    public void addInterceptors(InterceptorRegistry registry) {
        registry.addInterceptor(getInterceptor())
                //配置拦截器拦截所有的URL
                .addPathPatterns("/**")
                //配置拦截器不拦截以下URL
                .excludePathPatterns
                    ("/user/login","/user/register","/file/upload");
    }
}
```

注意：由于过滤器和拦截器作用相似，所以建议将 DemoApplication 启动类上的 @ServletComponentScan 注解注释掉，方便后续测试。

(3) 在 Postman 中创建一个新的请求，请求地址为 http://localhost:8080/user/login，响应结果为 ok，说明拦截器没有对此 URL 地址进行拦截，如图 9-4 所示。

(4) 将请求地址修改为 http://localhost:8080/user/getUserInfo，返回错误提示信息，说明这个错误被拦截器所拦截，如图 9-5 所示。

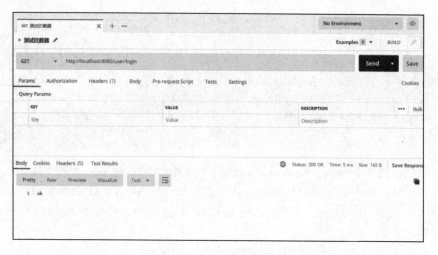

图 9-4 拦截器测试(1)

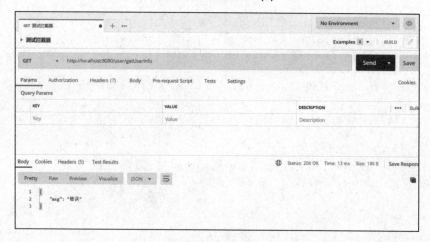

图 9-5 拦截器测试(2)

(5) 在 Headers 请求中增加 token 参数,发送请求,响应 ok,说明请求通过拦截器,如图 9-6 所示。

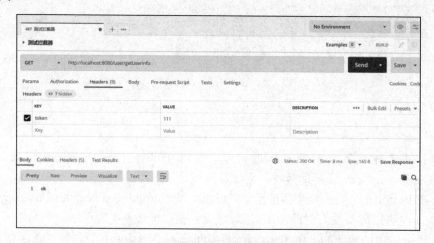

图 9-6 拦截器测试(3)

## 9.3 文件上传

动画：解密 SpringBoot 的文件上传：简单、快捷、灵活

微课：文件上传

SpringBoot 提供了简单而强大的文件上传功能，可以让开发人员轻松地处理文件上传和存储。文件上传在许多 Web 应用程序中都是常见的需求，例如上传图片、文档、音频文件等。

下面演示 SpringBoot 文件上传功能的使用。

（1）在 resources 目录下创建 application.yml 配置文件，设置上传文件的大小，其中 max-file-size 设置上传的单个文件大小的最大值，max-request-size 设置一次请求总的上传文件大小的最大值，代码如下：

```yaml
spring:
  servlet:
    multipart:
      max-file-size: 10MB
      max-request-size: 10MB
```

（2）在 com.demo.controller 包下创建 FileController 控制器，编写文件上传的方法，代码如下：

```java
package com.demo.controller;

import org.springframework.web.bind.annotation.PostMapping;
import org.springframework.web.bind.annotation.RequestMapping;
import org.springframework.web.bind.annotation.RestController;
import org.springframework.web.multipart.MultipartFile;

import java.io.File;
import java.io.IOException;

@RestController
@RequestMapping("file")
public class FileController {
    //设置文件存储的路径
    String path = "D:\\upload";

    @PostMapping("upload")
    public String upload(MultipartFile file) throws IOException {
        File uploadFilePath = new File(path);
        if(!uploadFilePath.exists()){
            uploadFilePath.mkdir();
        }
        //存储上传的文件
        file.transferTo(new File(uploadFilePath,file.getOriginalFilename()));
        return "ok";
    }

}
```

(3) 运行应用，在 Postman 下创建一个新的请求，地址为 http://localhost:8080/file/upload，在 Body 选项卡中选择 form-data，在下边输入框 KEY 里面输入 file，VALUE 选择一个本地文件，如图 9-7 所示。

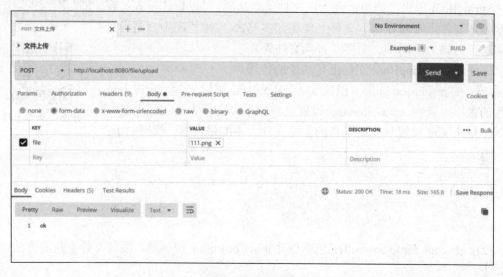

图 9-7　文件上传

(4) 在本地电脑 D 盘下找到 upload 目录，可以看到图片已经上传成功，如图 9-8 所示。

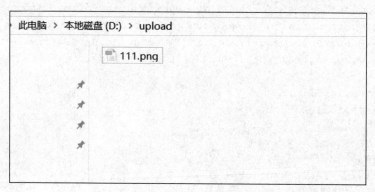

图 9-8　文件上传结果

## 9.4　SpringBoot 整合 Redis

### 9.4.1　Redis 简介

Redis(Remote Dictionary Server)是一个开源的，基于键值对的高性能数据存储系统。它通常被用作数据库、缓存和消息队列。Redis 支持多种数据结构，如字符串(String)、哈希(Hash)、列表(List)、集合(Set)、有序集合(Sorted Set)、位图(Bitmap)、地理空间(Geospatial)索引和流(Stream)等。

微课：Redis 缓存

Redis 的主要特点如下。

1. 高性能

Redis 能读的 QPS(每秒查询率)可达到每秒 11 万次，写的 QPS 可达到每秒 8.1 万次。

2. 持久化

Redis 提供了多种持久化方式，包括 RDB 快照和 AOF 日志文件，以保证数据的安全。

3. 支持事务

Redis 支持简单的事务功能。

4. 丰富的数据类型

Redis 除了支持简单的键值对外，还支持其他类型的数据结构。

5. 支持分布式

通过分片技术，Redis 可以很容易地扩展到多个服务器。

6. 丰富的客户端支持

大多数流行语言都有 Redis 客户端，包括 Java、Python、Ruby、Go 等。

7. 简单易用

Redis 有着非常简单的 API 和一个直观的命令行界面。

8. 社区活跃

由于 Redis 是开源软件，它有一个非常活跃的社区，这意味着不断的更新和大量的第三方库。

9. 灵活的配置和部署

可以嵌入式使用，也可以作为网络服务独立部署。

10. 主从复制和高可用

Redis 支持主从复制，提供高可用和数据冗余。

SpringBoot 与 Redis 的整合之所以流行，是因为它结合了 SpringBoot 的易用性和 Redis 的高性能特性，为现代应用程序提供了强大的数据处理和缓存解决方案。Redis 是一个内存中的数据结构存储，提供了极高的读写速度。这对于需要快速响应的 Web 应用来说是非常重要的。通过使用 Redis 作为缓存，可以显著提高应用程序的性能和用户体验。将频繁查询的数据缓存在 Redis 中，可以减少对数据库的访问次数，从而减轻数据库的负担，提高整体的应用性能。

## 9.4.2 添加 Redis Maven 依赖

若想使用 Redis，首先需要添加 Redis Maven 依赖，代码如下：

```
<!-- JUnit -->
<dependency>
```

```xml
        <groupId>org.springframework.boot</groupId>
        <artifactId>spring-boot-starter-test</artifactId>
        <scope>test</scope>
</dependency>

<!-- SpringBoot 整合 Redis -->
<dependency>
        <groupId>org.springframework.boot</groupId>
        <artifactId>spring-boot-starter-data-redis</artifactId>
</dependency>
```

### 9.4.3 下载与启动 Redis 服务端、客户端

下面将在 Windows 系统下讲解 Redis 的下载安装步骤。

(1) 下载 Redis。官方的 Redis 只支持 Linux 操作系统，在 Windows 系统上开发需要下载专用的 Windows 版本。访问链接 https://github.com/microsoftarchive/redis/releases/download/win-3.2.100/Redis-x64-3.2.100.zip。本书选用的是 Windows 系统的 3.2.10 版本。

(2) 解压下载 Windows 版本的 Redis 到指定的文件夹，如图 9-9 所示。

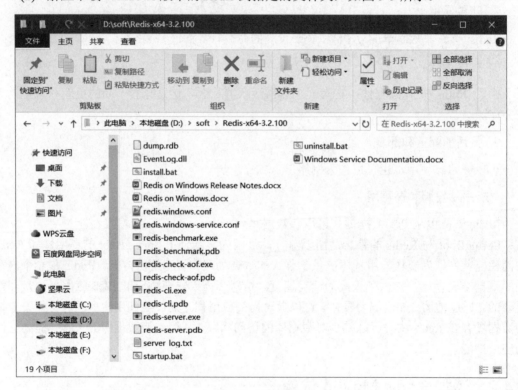

图 9-9 解压目录

(3) 在地址栏中输入 cmd，打开 cmd 命令行窗口，输入 "redis.server.exe redis.windows.conf"，启动 Redis 服务，启动后如图 9-10 所示。

(4) 双击 redis-cli.exe 启动 Redis 客户端，如图 9-11 所示。

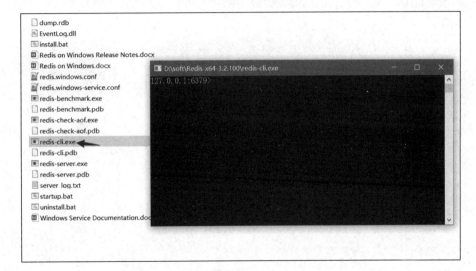

图 9-10 Redis 服务启动

图 9-11 Redis 客户端启动

## 9.4.4 编写 Redis 使用代码

在 src/main/java 目录下的 com.demo.service 包中创建 Redis 的使用代码类 MyRedisService。本例中演示对 Redis 存取字符串的功能，使用 StringRedisTemplate 类的 set 方法存储字符串，使用 get 方法获取字符串，具体代码如下：

```
package com.demo.service;

import org.springframework.beans.factory.annotation.Autowired;
import org.springframework.data.redis.core.StringRedisTemplate;
import org.springframework.stereotype.Service;

@Service
public class MyRedisService {
```

```java
//自动注入StringRedisTemplate类对象
@Autowired
private StringRedisTemplate stringRedisTemplate;
//向Redis中存储键的key值value
public void set(String key, String value) {
    stringRedisTemplate.opsForValue().set(key, value);
}
//从Redis中读取键的key值
public String get(String key) {
    return stringRedisTemplate.opsForValue().get(key);
}
}
```

### 9.4.5 单元测试

在src/test/java目录下的com.demo.test包中创建Redis的使用测试类MyRedisServiceTest，具体代码如下：

```java
package com.demo.test;

import com.demo.DemoApplication;
import com.demo.service.MyRedisService;
import org.junit.jupiter.api.Test;
import org.springframework.beans.factory.annotation.Autowired;
import org.springframework.boot.test.context.SpringBootTest;

@SpringBootTest(classes = DemoApplication.class)
public class MyRedisServiceTest {

    @Autowired
    private MyRedisService redisService;

    @Test
    public void test1(){
        redisService.set("username","zhangsan");
    }

    @Test
    public void test2(){
        String s = redisService.get("username");
        System.out.println(s);
    }
}
```

执行单元测试后，通过Redis客户端工具查看数据，数据"zhangsan"已经保存到了Redis中，如图9-12所示。

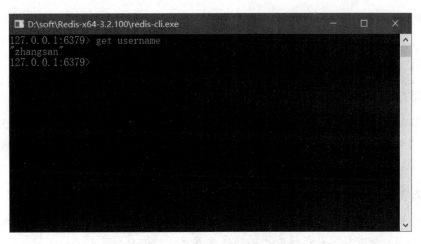

图 9-12　Redis 客户端查看结果

## 本 章 小 结

拦截器和过滤器都可以用于请求处理中的预处理和后处理，但它们有不同的应用场景和用途。选择合适的机制取决于需求。

文件上传是 Web 应用程序中常见的需求之一，SpringBoot 提供了便捷的方式来处理文件上传，但需要注意文件大小和类型的限制。

## 课 后 习 题

一、选择题

1. 在 SpringBoot 中，过滤器和拦截器的主要区别是(　　)。
   A. 过滤器可以修改 HTTP 请求和响应，拦截器只能修改请求
   B. 过滤器只能用于日志记录，拦截器可以用于身份验证和授权
   C. 过滤器在请求处理前后都可以介入，拦截器只能在处理器方法前后介入
   D. 过滤器只能用于处理 GET 请求，拦截器可以处理所有类型的请求
2. 在 SpringBoot 中，文件上传通常使用(　　)类来处理。
   A. FileInputStream　　B. MultipartFile　　C. FileReader　　D. Blob
3. 要创建一个自定义的 SpringBoot 拦截器，通常需要实现(　　)接口。
   A. Filter　　　　　　　　　　　　B. HandlerInterceptor
   C. HttpServletRequestWrapper　　　D. ResponseWrapper

二、填空题

1. 在 SpringBoot 中，要配置一个过滤器，通常需要创建一个实现 javax.servlet.Filter 接口的类，并使用_____注解将其标记为 Spring 组件。
2. 在 SpringBoot 中，要处理文件上传，通常需要使用_____类来接收上传的文件数据。

### 三、判断题

1. 过滤器和拦截器在SpringBoot中的使用场景和功能是完全相同的。（  ）
2. 文件上传通常涉及将文件数据保存到数据库中，而不是本地磁盘。（  ）

### 四、简答题

1. 请解释什么是SpringBoot中的过滤器和拦截器，以及它们的主要区别和使用场景。
2. 请描述在SpringBoot中处理文件上传的一般步骤，并示例说明如何实现文件上传功能。

### 五、实操题

创建一个包含文件上传功能的SpringBoot应用程序。用户应能够上传图像文件(例如JPEG或PNG)，并将其保存到服务器的指定目录中。

修改上述文件上传应用程序，添加一个拦截器来记录每个文件上传的时间戳。这个拦截器可以在文件上传前和文件保存后分别记录时间戳。

#### 信息过滤与个人观点的平衡

在使用过滤器和拦截器进行信息过滤时，我们不仅要关注信息安全，还需注意保护用户的个人观点和个人隐私，这在符合法律规定和维护社会稳定的同时，也体现了对人民当家做主的权利和社会主义法治精神的尊重。

# 第 10 章
# 项目实战：教学信息管理系统

**学习目标**

1. 了解基于 Spring、SpringMVC、SpringBoot、MyBatis 框架的教学信息管理系统的架构。
2. 熟悉系统开发环境搭建的详细步骤。
3. 掌握身份验证功能的开发。
4. 掌握系统各功能模块的开发。
5. 掌握跨域功能的配置。

**学习要点**

1. 分析系统功能并创建数据库。
2. 创建实体类、DAO、Service 和 Controller。
3. 开发权限验证拦截器。
4. 开发跨域访问与查询分页功能。

本章知识点结构如图 10-1 所示。

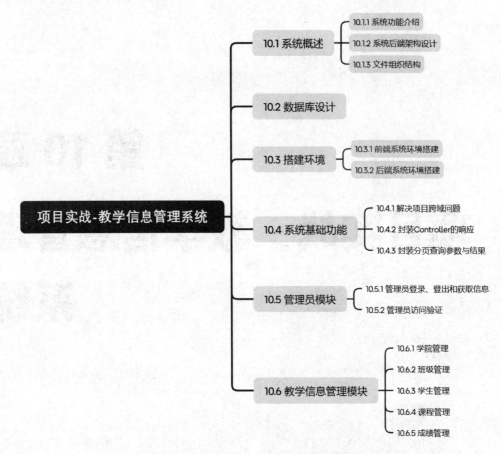

图 10-1　项目实战：教学信息管理系统

本章将使用前面各章节学习的 Spring、SpringMVC、SpringBoot、MyBatis 等框架来开发一个教学信息管理系统。

# 10.1　系统概述

## 10.1.1　系统功能介绍

本系统分为前端系统和后端系统两个子系统。后端系统使用本书介绍的 Spring、SpringMVC、SpringBoot 和 MyBatis 等框架编写，前端系统使用 Node.js、Vue.js 与 Element-Plus 等框架编写。

教学信息管理系统主要包含两大功能模块：管理员模块和教学信息管理模块。其中管理员模块实现了管理员的登录、登出和获取管理员信息功能；教学信息管理模块主要有学院、班级、学生、课程和成绩等信息的增删改查等功能。

系统后端的主要功能结构如图 10-2 所示。

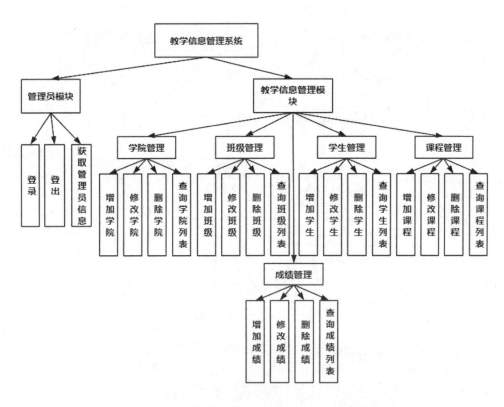

图 10-2　教学信息管理系统的主要功能结构

## 10.1.2　系统后端架构设计

系统后端架构设计主要包括以下部分。

(1) 持久层：该层由系统所涉及的实体类构成。

(2) 数据访问层(DAO 层)：主要使用 MyBatis 框架,该层由持久层的实体类对应的 DAO 接口和 MyBatis 映射文件组成,DAO 接口和映射文件的名称统一以 Dao 结尾。

(3) 业务逻辑层(Service 层)：主要实现系统的业务逻辑,该层主要由实体类对应的 Service 接口和其实现类组成，Service 接口以 Service 结尾,实现类以 ServiceImpl 结尾。

(4) 请求处理层：主要使用 SpringMVC 框架,包含 SpringMVC 的 Controller 类来接受系统前端页面发送的 Web 请求,调用业务逻辑层的 Service 实现类处理请求并将结果返回给前端页面。

系统使用 SpringBoot 框架作为基础结构，使用 Spring 框架来管理所有的对象的生命周期。

## 10.1.3　文件组织结构

在正式讲解系统的功能代码之前，先来了解一下系统中所涉及到的类、依赖、配置类、配置文件等文件在系统中的组织结构，如图 10-3 所示。

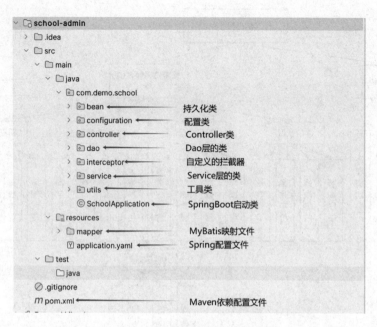

图 10-3　后端系统文件的组织结构

## 10.2　数据库设计

本系统后端包含管理员模块和教学信息管理模块，涉及到的实体有管理员、学院、班级、学生、课程和成绩。其中学生隶属于班级，班级隶属于学院，成绩和课程、学生有关联关系。系统的 ER 图如图 10-4 所示。

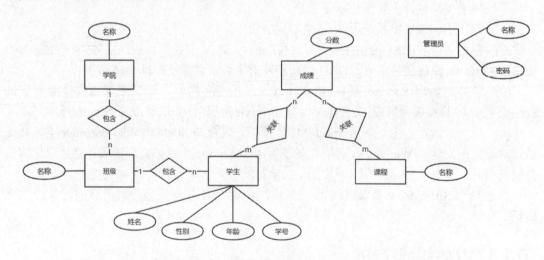

图 10-4　教学信息管理系统 ER 图

根据以上 ER 图进行分析，系统共有管理员表、学院表、班级表、学生表、课程表和成绩表。该系统中各表的具体结构分别如表 10-1～表 10-5 所示。

表 10-1 管理员表(admin)的结构

| 字 段 名 | 类 型 | 长 度 | 是否主键 | 说 明 |
|---|---|---|---|---|
| id | int | 11 | 是 | 管理员 id |
| name | varchar | 45 | 否 | 管理员名 |
| password | varchar | 45 | 否 | 管理员密码 |
| token | varchar | 32 | 否 | 认证 token |
| expired_time | datetime | 0 | 否 | token 过期时间 |

表 10-2 学院表(college)的结构

| 字 段 名 | 类 型 | 长 度 | 是否主键 | 说 明 |
|---|---|---|---|---|
| id | int | 11 | 是 | 学院 id |
| name | varchar | 45 | 否 | 学院名称 |

表 10-3 班级表(class)的结构

| 字 段 名 | 类 型 | 长 度 | 是否主键 | 说 明 |
|---|---|---|---|---|
| id | int | 11 | 是 | 班级 id |
| name | varchar | 45 | 否 | 班级名称 |
| college_id | int | 11 | 否 | 隶属学院的 id |

表 10-4 学生表(student)的结构

| 字 段 名 | 类 型 | 长 度 | 是否主键 | 说 明 |
|---|---|---|---|---|
| id | int | 11 | 是 | 学生 id |
| name | varchar | 45 | 否 | 学生姓名 |
| serial_no | varchar | 45 | 否 | 学生学号 |
| gender | varchar | 45 | 否 | 性别 |
| age | int | 11 | 否 | 年龄 |
| class_id | int | 11 | 否 | 隶属班级的 id |

表 10-4 课程表(course)的结构

| 字 段 名 | 类 型 | 长 度 | 是否主键 | 说 明 |
|---|---|---|---|---|
| id | int | 11 | 是 | 课程 id |
| name | varchar | 45 | 否 | 课程名称 |

表 10-5 成绩表(score)的结构

| 字 段 名 | 类 型 | 长 度 | 是否主键 | 说 明 |
|---|---|---|---|---|
| id | int | 11 | 是 | id |
| course_id | int | 11 | 否 | 关联课程表的 id |
| student_id | int | 11 | 否 | 关联学生表的 id |
| score | int | 11 | 否 | 分数 |

## 10.3 搭建环境

### 10.3.1 前端系统环境搭建

由于本书主要的内容是讲解后端 Web 开发,所以前端系统开发相关技术不是本书的重点内容,同时由于篇幅有限,前端系统在本书中直接给出页面代码,其具体的架构设计和开发步骤将略过,读者如有兴趣可以阅读 Node.js、Vue.js、Element-Plus 等前端技术的相关图书或文档获取详细的信息,读者可以在本书提供的源代码中找到前端系统的完整代码。

开发前端系统需要使用 Node.js 工具,下面演示 Node.js 的安装过程。

**1. 安装 Node.js**

(1) 本书中使用的 Node.js 版本为 node-v18.17.1-x64,用户可以访问链接 https://nodejs.org/dist/v18.17.1/node-v18.17.1-x64.msi 下载安装包。

(2) 下载 Node.js 安装包后,双击安装文件,单击 Next 按钮,如图 10-5 所示。

(3) 选中接受软件使用条款复选框,单击 Next 按钮,如图 10-6 所示。

图 10-5 Node.js 的欢迎安装界面

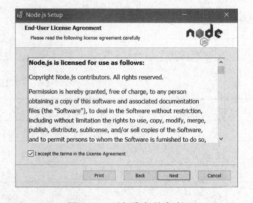

图 10-6 接受安装条款

(4) 选择软件安装路径,单击 Next 按钮,如图 10-7 所示。

(5) 选择默认安装,单击 Next 按钮,如图 10-8 所示。

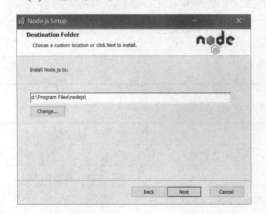

图 10-7 选择软件安装路径

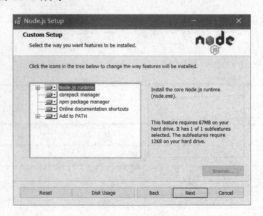

图 10-8 选择默认安装

(6) 单击 Next 按钮，如图 10-9 所示。

(7) 单击 Install 按钮，如图 10-10 所示。

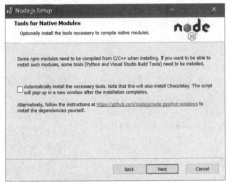
图 10-9　本机模块安装

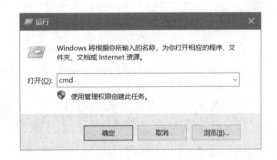
图 10-10　开始安装

(8) 单击 Finish 按钮，如图 10-11 所示。

(9) 在 Windows 系统中使用 Win+R 快捷键打开"运行"对话框，输入"cmd"并单击"确定"按钮，如图 10-12 所示。

图 10-11　安装结束

图 10-12　"运行"对话框

(10) 在命令行界面输入以下的命令并运行，设置 npm 的下载源为国内的镜像，如图 10-13 所示。

```
npm config set registry https://registry.npmmirror.com
```

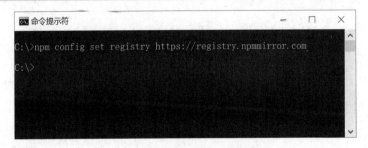
图 10-13　Node 环境配置

### 2. 创建前端项目

(1) 初始化项目，项目名称为 school-admin-ui，在系统命令行界面中依次执行下列命令：

```
npm init vite@latest school-admin-ui -- --template vue
cd school-admin-ui
npm install
```

(2) 安装 vue-router,在系统命令行界面中执行如下命令:

```
npm install vue-router@4
```

安装完成后在 school-admin-ui 的 src/router 文件夹下创建文件 index.js,文件内容如下:

```js
// 路由文件
import { createRouter, createWebHistory } from "vue-router";

const routes = []
const router = createRouter({
  history: createWebHistory(),
  routes
})
export default router;
```

然后在 src 文件夹下的 main.js 文件中引入上一步的 router,添加完成后 main.js 的文件内容如下:

```js
import { createApp } from 'vue'
import './style.css'
import App from './App.vue'
import Router from './router/index.js'

const app = createApp(App)
app.use(Router).mount('#app');
```

(3) 安装基于 Vue.js 的页面组件库 Element-Plus 和相关工具,在系统命令行界面中依次执行如下命令:

```
npm install element-plus
npm install -D unplugin-auto-import
npm install -D unplugin-vue-components
npm install -D unplugin-element-plus
npm install -D sass-embedded
```

(4) 安装 Element-Plus 的图标库,在系统命令行界面中执行如下命令:

```
npm install @element-plus/icons-vue
```

(5) 安装前端网络请求库 axios,在系统命令行界面中执行如下命令:

```
npm install axios
```

安装完成后在 src/utils 文件夹下创建文件 index.js,添加如下代码:

```js
export function localGet (key) {
  const value = window.localStorage.getItem(key)
  try {
    return JSON.parse(window.localStorage.getItem(key))
```

```
  } catch (error) {
    return value
  }
}
export function localSet (key, value) {
  window.localStorage.setItem(key, JSON.stringify(value))
}
export function localRemove (key) {
  window.localStorage.removeItem(key)
}
```

然后在 utils 文件夹下创建文件 axios.js,添加如下代码:

```
import axios from 'axios'
import config from '~/config'
import { localGet, localRemove } from './index'
import router from '@/router/index'

axios.defaults.baseURL = 'http://localhost:8080'

axios.defaults.headers['X-Requested-With'] = 'XMLHttpRequest'
axios.defaults.headers['Authorization'] = localGet('token') || ''
axios.defaults.headers.post['Content-Type'] = 'application/json'

axios.interceptors.response.use(res => {
  if (typeof res.data !== 'object') {
    ElMessage.error('服务端异常!')
    return Promise.reject(res)
  }
  if (res.data.responseCode != 200) {
    if (res.data.message) {
      ElMessage.error(res.data.message)
    }
    if (res.data.responseCode == 401) {
      localRemove("token")
      router.push({ path: '/login' })
    }
    return Promise.reject(res)
  }
  return res
})

export default axios
```

(6) 修改前端系统设置。

修改 school-admin-ui 文件夹下的 vite.config.js 文件,修改后的代码如下:

```
import { defineConfig } from 'vite'
import path from 'path'

import vue from '@vitejs/plugin-vue'
import AutoImport from 'unplugin-auto-import/vite'
import Components from 'unplugin-vue-components/vite'
```

```
import { ElementPlusResolver } from 'unplugin-vue-components/resolvers'
import ElementPlus from 'unplugin-element-plus/vite'
export default defineConfig({
  plugins: [
    vue(),
    AutoImport({
      resolvers: [ElementPlusResolver()],
    }),
    Components({
      resolvers: [ElementPlusResolver()],
    }),
    ElementPlus({
      useSource: true
    })
  ],
  resolve: {
    alias: {
      '~': path.resolve(__dirname, './'),
      '@': path.resolve(__dirname, 'src')
    },
  },
  base: './',
  server: {
    open: true,
  }
})
```

通过以上步骤，前端系统环境就已经搭建完毕，后续的开发过程中将陆续添加各功能页面。

## 10.3.2 后端系统环境搭建

### 1. 初始化数据库

初始化数据库的步骤如下。

(1) 在 Navicat 工具中，单击"查询"按钮，再单击"新建查询"按钮，如图 10-14 所示。

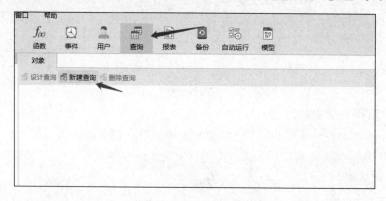

图 10-14 Navicat 界面

(2) 在新建查询中，执行下面的 SQL 语句创建数据库、数据表和插入一些初始测试数据：

```sql
-- 创建一个新的数据库
CREATE DATABASE if not exists school_admin DEFAULT CHARACTER SET utf8;

-- 切换到 school_admin
use school_admin;

-- 创建管理员表 admin
DROP TABLE IF EXISTS `admin`;
CREATE TABLE `admin` (
  `id` int NOT NULL AUTO_INCREMENT,
  `name` varchar(45) NOT NULL,
  `password` varchar(45) NOT NULL,
  `token` varchar(32) DEFAULT NULL,
  `expired_time` datetime DEFAULT NULL,
  PRIMARY KEY (`id`),
  UNIQUE KEY `name_UNIQUE` (`name`)
) ENGINE=InnoDB DEFAULT CHARSET=utf8mb4 COLLATE=utf8mb4_0900_ai_ci;
-- 插入初始管理员 admin 密码为 123456
INSERT INTO `admin`(name,password) VALUES('admin',md5('123456'));

-- 创建学院表
DROP TABLE IF EXISTS `college`;
CREATE TABLE `college` (
  `id` int NOT NULL AUTO_INCREMENT,
  `name` varchar(50) NOT NULL,
  PRIMARY KEY (`id`)
) ENGINE=InnoDB DEFAULT CHARSET=utf8mb4 COLLATE=utf8mb4_0900_ai_ci;
-- 插入学院表测试数据
INSERT INTO `college`(name) VALUES ('测试学院1'),('测试学院2'),('测试学院3'),('测试学院4'),('测试学院5');

-- 创建班级表
DROP TABLE IF EXISTS `class`;
CREATE TABLE `class` (
  `id` int NOT NULL AUTO_INCREMENT,
  `name` varchar(45) DEFAULT NULL,
  `college_id` int DEFAULT NULL,
  PRIMARY KEY (`id`),
  KEY `fk_class_college_idx` (`college_id`),
  CONSTRAINT `fk_class_college` FOREIGN KEY (`college_id`) REFERENCES `college` (`id`)
) ENGINE=InnoDB DEFAULT CHARSET=utf8mb4 COLLATE=utf8mb4_0900_ai_ci;
-- 插入班级表测试数据
INSERT INTO `class`(name,college_id) VALUES ('班级1',1), ('班级2',2), ('班级3',3), ('班级4',4), ('班级5',5);
-- 创建学生表
DROP TABLE IF EXISTS `student`;
CREATE TABLE `student` (
  `id` int NOT NULL AUTO_INCREMENT,
  `name` varchar(45) NOT NULL,
```

```sql
  `serial_no` varchar(45) NOT NULL,
  `gender` varchar(45) DEFAULT NULL,
  `age` int DEFAULT NULL,
  `class_id` int DEFAULT NULL,
  PRIMARY KEY (`id`),
  UNIQUE KEY `serial_no_UNIQUE` (`serial_no`),
  UNIQUE KEY `name_UNIQUE` (`name`),
  KEY `fk_student_class_idx` (`class_id`),
  CONSTRAINT `fk_student_class` FOREIGN KEY (`class_id`) REFERENCES `class` (`id`)
) ENGINE=InnoDB DEFAULT CHARSET=utf8mb4 COLLATE=utf8mb4_0900_ai_ci;
-- 插入学生表测试数据
INSERT INTO `student`(name,serial_no,gender,age,class_id) VALUES ('测试学生 1','20240101','male',18,1),('测试学生 2','20240102','male',21,2),('测试学生 3','20240103','male',21,3),('测试学生 4','20240104','female',22,4),('测试学生 5','20240105','male',22,5),('测试学生 6','20240106','male',22,1),('测试学生 7','20240107','male',21,2),('测试学生 8','20240108','female',19,3),('测试学生 9','20240109','male',22,4) ,('测试学生 10','20240110','female',20,5);
-- 创建课程表
DROP TABLE IF EXISTS `course`;
CREATE TABLE `course` (
  `id` int NOT NULL AUTO_INCREMENT,
  `name` varchar(45) NOT NULL,
  PRIMARY KEY (`id`)
) ENGINE=InnoDB DEFAULT CHARSET=utf8mb4 COLLATE=utf8mb4_0900_ai_ci;
-- 插入课程表测试数据
INSERT INTO `course`(name) VALUES ('Java 服务端开发'),('开源大数据基础'),('数据仓库原理与应用'),('思想道德与法治'),('形势与政策'),('数据可视化'),('云计算平台运维与开发'),('大数据分析与应用'),('大数据采集与处理'),('Java 语言程序设计'),('Python 语言程序设计');
-- 创建成绩表
DROP TABLE IF EXISTS `score`;
CREATE TABLE `score` (
  `id` int NOT NULL AUTO_INCREMENT,
  `student_id` int NOT NULL,
  `course_id` int NOT NULL,
  `score` int NOT NULL,
  PRIMARY KEY (`id`),
  KEY `fk_score_course_idx` (`course_id`),
  KEY `fk_score_student_idx` (`student_id`),
  CONSTRAINT `fk_score_course` FOREIGN KEY (`course_id`) REFERENCES `course` (`id`),
  CONSTRAINT `fk_score_student` FOREIGN KEY (`student_id`) REFERENCES `student` (`id`)
) ENGINE=InnoDB DEFAULT CHARSET=utf8mb4 COLLATE=utf8mb4_0900_ai_ci;
```

### 2. 创建项目，添加依赖

在 IntelliJ IDEA 中创建一个新的 Maven 项目，项目名为 school-admin。在 pom.xml 文件中添加依赖。系统需要引入 SpringBoot、Spring、SpringMVC、MyBatis、MySQL 等依赖项，其中 spring-boot-starter-parent 会将 SpringBoot、Spring 等依赖项引入，spring-boot-starter-web 引入 SpringMVC 所需的依赖项，mybatis-spring-boot-starter 引入 MyBatis 及其与

SpringBoot 整合的依赖项，mysql-connector-java 引入 MySQL 的 Java 驱动依赖项。代码如下：

```xml
<?xml version="1.0" encoding="UTF-8"?>
<project xmlns="http://maven.apache.org/POM/4.0.0"
xmlns:xsi="http://www.w3.org/2001/XMLSchema-instance"
    xsi:schemaLocation="http://maven.apache.org/POM/4.0.0
https://maven.apache.org/xsd/maven-4.0.0.xsd">
    <modelVersion>4.0.0</modelVersion>
<!--spring-boot-starter-parent 会将 SpringBoot、Spring 等依赖项引入-->
    <parent>
        <groupId>org.springframework.boot</groupId>
        <artifactId>spring-boot-starter-parent</artifactId>
        <version>2.7.9</version>
    </parent>
    <groupId>com.biem</groupId>
    <artifactId>school-admin</artifactId>
    <version>0.0.1-SNAPSHOT</version>
    <name>school-admin</name>
    <description>school-admin</description>
    <properties>
        <java.version>1.8</java.version>
    </properties>
    <dependencies>
<!--引入 SpringMVC 所需的依赖项-->
        <dependency>
            <groupId>org.springframework.boot</groupId>
            <artifactId>spring-boot-starter-web</artifactId>
        </dependency>
<!--引入 MySQL 的 Java 驱动依赖项-->
        <dependency>
            <groupId>mysql</groupId>
            <artifactId>mysql-connector-java</artifactId>
            <version>5.1.49</version>
        </dependency>
<!--引入 MyBatis 及其与 SpringBoot 整合的依赖项-->
        <dependency>
            <groupId>org.mybatis.spring.boot</groupId>
            <artifactId>mybatis-spring-boot-starter</artifactId>
            <version>2.1.3</version>
        </dependency>
<!--引入一些工具类的依赖项-->
        <dependency>
            <groupId>org.apache.commons</groupId>
            <artifactId>commons-lang3</artifactId>
            <version>3.11</version>
        </dependency>
    </dependencies>
    <build>
        <plugins>
<!--SpringBoot 程序打包插件-->
            <plugin>
```

```xml
            <groupId>org.springframework.boot</groupId>
            <artifactId>spring-boot-maven-plugin</artifactId>
            <version>2.6.3</version>
        </plugin>

    </plugins>
  </build>
</project>
```

### 3. 创建 Spring 配置文件 application.yaml

在 resources 目录下创建 application.yaml 配置文件，添加项目配置。配置内容主要包含 Spring 框架配置、MyBatis 配置和服务器配置三个部分，这些配置在前面章节中已有介绍。完整的 application.yaml 文件代码如下：

```yaml
# Spring 配置项
spring:
# 数据库信息
  datasource:
    url: jdbc:mysql://localhost:3306/school_admin
    username: root
    password: 1234
    driver-class-name: com.mysql.cj.jdbc.Driver
# MyBatis 配置项
mybatis:
  configuration:
    #自动将数据库下划线命名转换为驼峰命名
    map-underscore-to-camel-case: true
    #指定 MyBatis 映射文件的位置为 mapper 文件夹下所有以 Dao 结尾的 xml 文件
  mapper-locations: classpath:mapper/*Dao.xml
#设置 MyBatis 自动扫描此项中定义的所有包下的类并将其注册到 MyBatis 的别名列表中
#之后在 MyBatis 的 mapper 文件中使用这些类的时候只需要写类的名称，可以省去其完整的包名
  type-aliases-package: com.demo.school.bean,com.demo.school.utils
server:
# 设置 SpringBoot 启动服务访问端口
  port: 8080
```

### 4. 创建项目所需的包和类

（1）在项目的 src/main/java 目录下，创建 com.demo.school 包，然后在此包下依次创建项目所需的各个子包，分别为 bean、configuration、controller、dao、interceptor、service 和 utils。在 service 包中创建子包 impl。

（2）在项目的 src/main/resources 目录下创建 mapper 文件夹。

（3）在项目的 src/main/java 目录下的 com.demo.school 包中创建 SpringBoot 项目的启动类 SchoolApplication，代码如下：

```java
package com.demo.school;

import org.springframework.boot.SpringApplication;
```

```
import org.springframework.boot.autoconfigure.SpringBootApplication;

@SpringBootApplication
public class SchoolApplication {

    public static void main(String[] args) {
        SpringApplication.run(SchoolApplication.class, args);
    }

}
```

## 10.4 系统基础功能

在本章以上内容中已经完成了教学信息管理系统的前端系统和后端系统的项目环境搭建，在开始(如管理员登录和教学信息管理功能)开发之前，先来开发项目所需的基础功能，解决前端和后端访问之间和跨域问题、封装后端的 Controller 的响应和封装分页功能。

### 10.4.1 解决项目跨域问题

教学信息管理系统的前端和后端是分离的两个系统，两个系统会在不同的地址和端口上启动，当用户使用前端操作时，前端页面会向后端系统发送 HTTP 请求，就会产生跨域问题。所谓跨域问题就是浏览器为了保护用户的信息安全，实施了同源策略(Same-Origin Policy)，即只允许页面请求同源(相同协议、域名和端口)的资源，当 JavaScript 发起的请求跨越了同源策略，即请求的目标与当前页面的域名、端口、协议不一致时，浏览器会阻止请求的发送或接收。例如，本系统的前端地址是 http://127.0.0.1:5173，而后端的访问地址是 http://127.0.0.1:8080，它们的访问地址域名相同但端口不同，必然会出现跨域问题，浏览器将会报类似下面的错误：

Access to XMLHttpRequest at '......' from origin '......' has been blocked by CORS policy: No 'Access-Control-Allow-Origin' header is present on the requested resource。

在 SpringBoot 中提供了解决此问题的全局配置方法，即重写 WebMvcConfigurer 接口的 addCorsMappings 方法来实现跨域。本系统的开发中将采用此方法。

在项目的 src/main/java 目录下的 com.demo.school.configuration 包中，创建 SchoolConfiguration 类，代码如下：

```
package com.demo.school.configuration;

import org.springframework.boot.SpringBootConfiguration;
import org.springframework.web.servlet.config.annotation.CorsRegistry;
import org.springframework.web.servlet.config.annotation.WebMvcConfigurer;

@SpringBootConfiguration
public class SchoolConfiguration implements WebMvcConfigurer {
    public void addCorsMappings(CorsRegistry registry) {
        registry.addMapping("/**")      //添加映射路径，"/**"表示对所有的路径进行全局跨域访问权限的设置
```

```
            .allowedOriginPatterns("*")  //开放哪些 ip、端口、域名的访问权限
            .allowCredentials(true)      //是否允许发送 Cookie 信息
            .allowedMethods("GET", "POST", "PUT", "DELETE")
                //开放哪些 HTTP 方法，允许跨域访问
            .allowedHeaders("*")         //允许 HTTP 请求中所有的 Header 信息
            .exposedHeaders("*");
              //暴露哪些头部信息(因为跨域访问默认不能获取全部头部信息)
        }
    }
```

在上面的代码中，SchoolConfiguration 类实现了 WebMvcConfigurer 接口并重写了 addCorsMappings 方法，在方法中为了简便起见，对所有的配置都采用了最宽松的限制，其各个方法的作用如下。

(1) addMapping：设置后端系统允许跨域的路径，"/**"表示后端系统所有的路径都允许跨域访问。

(2) allowedOriginPatterns：设置后端系统允许跨域的请求来源，"*"表示后端系统允许任何的来源地址跨域访问。

(3) allowCredentials：当为 true 时，允许浏览器在发起跨域请求时携带认证信息(如 cookies)。

(4) allowedMethods：指定哪些 HTTP 方法可以被用于跨域请求，"GET"、"POST"、"PUT"、"DELETE"表示后端系统允许前端使用 HTTP 的 GET、POST、PUT 和 DELETE 4 种请求方法跨域访问。

(5) allowedHeaders：设置允许前端发送的请求列表，"*"表示后端系统允许任何请求头。

(6) exposedHeaders：设置后端可以返回哪些请求头给前端的跨域请求，"*"表示后端可以返回任何请求头。

经过以上配置之后，前端页面到后端的请求将可以顺利访问。

## 10.4.2 封装 Controller 的响应

在本书前面的章节中开发 Controller 时，对于返回的响应并没有做统一的封装处理，如 6.2.3 节中 BookController 的方法中，返回值有 Boolean 型、Book 型和 List<Book>等多种不同的类型。在实际的前后端分离的系统开发中，因为前端系统需要访问后端系统 Controller 中的 Web 服务并解析其响应然后展示页面，如果每个后端服务的响应都是不同的类型显然不利于前端代码的开发。因此需要对后端 Controller 中的响应做统一的封装，使其返回同样的类型，这样前端系统就能够方便进行统一的处理。

下面来定义响应结果需要的信息。

首先响应结果需要有信息表示其是成功结果还是失败结果，一般设置两个信息项：一个是响应代码，一个是响应信息。响应代码采用与 HTTP 响应代码相似的规范，200 为成功结果，400 和 500 为失败结果。响应信息如果成功为 SUCCESS，失败则是具体的失败原因信息。然后是响应的具体数据，对于一些请求的响应只有成功或失败信息，如数据插入、数据更新和数据删除等，而对于另一些查询类请求，则还需要加上查询的数据集结果，如查询单个课程的信息、查询课程列表信息等。

根据以上分析，下面开始进行封装响应代码的开发。

(1) 在项目的 src/main/java 目录下的 com.demo.school.utils 包中，创建 Response 类，代码如下：

```java
package com.demo.school.utils;
import java.io.Serializable;
public class Response<T> implements Serializable {
    private static final long serialVersionUID = 1L;
    //响应代码
    private int responseCode;
    //响应信息
    private String message;
    //响应数据集
    private T data;
    //构造方法
    public Response() {
    }
    public Response(int responseCode, String message) {
        this.responseCode = responseCode;
        this.message = message;
    }
    // getter 和 setter 方法省略
}
```

int 类型的 responseCode 为响应的代码，String 类型 message 为响应信息，data 为返回的数据集结果，这里使用泛型。

(2) 为了更加方便的调用，下面创建一个返回 Response 的工厂类。在项目的 src/main/java 目录的下 com.demo.school.utils 包中，创建 ResultGenerator 类，代码如下：

```java
package com.demo.school.utils;

import org.springframework.util.StringUtils;

public class ResultGenerator {
    //成功响应信息
    public static final String MESSAGE_SUCCESS = "SUCCESS";
    //失败响应信息
    public static final String MESSAGE_FAILURE = "FAILURE";
    //成功响应代码
    public static final int RESULT_CODE_SUCCESS = 200;
    //失败响应代码
    public static final int RESULT_CODE_BAD_REQUEST_ERROR = 400;
    //失败响应代码
    public static final int RESULT_CODE_UNAUTHORIZED_ERROR = 401;
    //失败响应代码
    public static final int RESULT_CODE_SERVER_ERROR = 500;

    //创建数据集为空的成功响应
    public static Response genSuccessResult() {
        Response response = new Response();
        response.setResponseCode(RESULT_CODE_SUCCESS);
```

```java
        response.setMessage(MESSAGE_SUCCESS);
        return response;
    }

    //创建可以自定义信息的成功响应
    public static Response genSuccessResultMessage(String message) {
        Response response = new Response();
        response.setResponseCode(RESULT_CODE_SUCCESS);
        response.setMessage(message);
        return response;
    }

    //创建可以自定义数据集的成功响应
    public static Response genSuccessResultData(Object data) {
        Response response = new Response();
        response.setResponseCode(RESULT_CODE_SUCCESS);
        response.setMessage(MESSAGE_SUCCESS);
        response.setData(data);
        return response;
    }

    //创建可以自定义信息的失败响应
    public static Response genFailResult(String message) {
        Response response = new Response();
        response.setResponseCode(RESULT_CODE_SERVER_ERROR);
        if (StringUtils.hasLength(message) && StringUtils.hasText(message)) {
            response.setMessage(MESSAGE_FAILURE);
        } else {
            response.setMessage(message);
        }
        return response;
    }

    //创建可以自定义代码和信息的失败响应
    public static Response genFailResult(int code, String message) {
        Response response = new Response();
        response.setResponseCode(code);
        response.setMessage(message);
        return response;
    }
}
```

在 ResultGenerator 类中已经定义好了能够返回各类响应结果的方法，在后续的功能模块开发中可以直接调用。

## 10.4.3 封装分页查询参数与结果

在系统的功能介绍中提到教学信息管理涉及到各项信息列表的查询，包含查询学院、班级、学生、课程、成绩列表。当数据库中的信息较多时，一次性返回所有数据给前端并且显示出来，这种方式会导致两个问题。一是一次返回的数据过多会影响数据库的性能，也会影响页面的响应速度；二是一页中显示大量数据不方便用户浏览和查找。因此，分页功能是各种信息管理系统中不可缺少的一项。

本节先将分页功能需要的请求参数和返回结果格式做统一封装，在后续开发各项具体查询信息列表功能部分会应用本节的封装实现分页功能。

1. 封装分页请求参数

前端系统向后端系统发送分页查询时，必不可少的参数有两个：请求的页码和每页显示的数据条数。由请求页码和每页数据条数可以计算出当前页的第一条数据在总的数据集中的位置。现在用 pageNumber 表示页码，pageSize 表示每页数据条数，start 表示第一条数据的索引。在分页查询时使用 MySQL 的 LIMIT 功能，分页 LIMIT 的语法格式如下：

```
SELECT * FROM 表名 LIMIT #{start},#{pageSize};
```

其中，在 MySQL 中 start 参数是从 0 开始，所以 start 参数的计算方式为：start=(pageNumber - 1) * pageSize。

根据以上的内容，下面创建分页查询请求的封装类，在 school-admin 文件夹的 src/main/java 目录下的 com.demo.school.utils 包中创建分页查询类 PageQuery，代码如下：

```java
package com.demo.school.utils;

public class PageQuery {
    //当前页码
    private int pageNumber;
    //每页条数
    private int pageSize;
    // 当前页的第一条记录在总数据中的索引,由 pageNumber 和 pageSize 计算得出
    private int start;
    public PageQuery(int pageNumber, int pageSize) {
        this.pageNumber = pageNumber;
        this.pageSize = pageSize;
        //计算出 start 的值
        this.start = (pageNumber - 1) * pageSize;
    }
// getter 和 setter 方法省略
}
```

2. 封装分页查询结果

前端系统显示分页需要的信息有数据总数、数据总页数、每页数据数、当前页码和当前页的数据集结果。

在 school-admin 文件夹的 src/main/java 目录下的 com.demo.school.utils 包中创建分页查询结果类 PageResponse，代码如下：

```java
package com.demo.school.utils;

import java.util.List;

public class PageResponse<T>{
//数据总数
```

```java
    private int total;
    //每页数据数
    private int pageSize;
    //总页数
    private int totalPage;
    //当前页码
    private int currentPage;
    //当前页数据集结果
    private List<T> list;
    //构造方法
    public PageResponse(List<T> list, int total, int pageSize, int currentPage)
{
        this.list = list;
        this.total = total;
        this.pageSize = pageSize;
        this.currentPage = currentPage;
        this.totalPage = (int) Math.ceil((double) total / pageSize);
    }
//getter 和 setter 方法省略
}
```

以上两部分就完成了分页功能的封装，在开发教学信息管理模块时可以使用封装类完成分页查询的开发。

## 10.5 管理员模块

管理员模块包含管理员登录、登出和获取管理员信息并在前端页面显示三项。除此之外，后端系统还需要添加一个拦截器用来拦截所有的请求并检查请求是否带有正确的认证信息。

### 10.5.1 管理员登录、登出和获取信息

管理员登录的主要流程如图 10-15 所示。

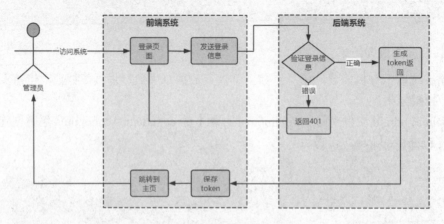

图 10-15　管理员登录流程

由图 10-15 可知，管理员登录流程具体如下。

（1）管理员访问前端系统登录页面，输入用户名和密码，单击登录后前端向后端发送登录请求。

（2）后端系统接收到登录信息后检查用户名和密码是否正确。

（3）如果登录信息正确则生成 token 保存在数据库的 admin 表中并返回给前端系统，前端系统保存 token 以便在随后的所有请求中都带上 token 访问。

（4）如果登录信息错误则返回代码为 401 的响应，前端系统获取 401 响应认定登录失败，仍停留在登录页面并在页面显示登录失败信息。

管理员的登出功能相对登录较为简单，就是将数据表 admin 中对应的 token 信息删除即可。

管理员信息获取是从数据表 admin 中查询发出请求的管理员名称并显示在前端系统的页面上。

在开发管理员功能时，先开发前端系统，然后开发后端系统。在 10.3.1 节已经提到，本书主要是讲解后端开发的技术，所以前端系统方面的技术不做详细讲解，直接给出代码。

1. 开发前端

前端系统在 10.3.1 节中进行了初始化，下面的步骤在前端系统 school-admin-ui 中进行。

（1）在 school-admin-ui 的 src 文件夹下找到 App.vue 文件，App.vue 文件为前端系统页面的主文件，编辑文件代码如下：

```
<template>
  <div class="layout">
    <el-container v-if="state.showMenu" class="container">
      <el-aside class="aside">
        <div class="head">
          <div>
            <span>教学信息管理系统</span>
          </div>
        </div>
        <div class="line" />
        <el-menu background-color="#222832" text-color="#fff" :unique-opened="true" :router="true" :default-active="state.currentPath">
          <el-sub-menu index="1">
            <template #title>
              <span>学院管理</span>
            </template>
            <el-menu-item-group>
              <el-menu-item index="/college"><el-icon>
                <Grid />
              </el-icon>学院列表</el-menu-item>
              <el-menu-item index="/addCollege"><el-icon>
                <FullScreen />
              </el-icon>添加学院</el-menu-item>
            </el-menu-item-group>
          </el-sub-menu>
          <el-sub-menu index="2">
            <template #title>
```

```html
        <span>班级管理</span>
      </template>
      <el-menu-item-group>
        <el-menu-item index="/class"><el-icon>
            <Grid />
          </el-icon>班级列表</el-menu-item>
        <el-menu-item index="/addClass"><el-icon>
            <House />
          </el-icon>添加班级</el-menu-item>
      </el-menu-item-group>
    </el-sub-menu>
    <el-sub-menu index="3">
      <template #title>
        <span>学生管理</span>
      </template>
      <el-menu-item-group>
        <el-menu-item index="/student"><el-icon>
            <Grid />
          </el-icon>学生列表</el-menu-item>
        <el-menu-item index="/addStudent"><el-icon>
            <UserFilled />
          </el-icon>添加学生</el-menu-item>
      </el-menu-item-group>
    </el-sub-menu>
    <el-sub-menu index="4">
      <template #title>
        <span>课程管理</span>
      </template>
      <el-menu-item-group>
        <el-menu-item index="/course"><el-icon>
            <Menu />
          </el-icon>课程列表</el-menu-item>
        <el-menu-item index="/addCourse"><el-icon>
            <Goods />
          </el-icon>添加课程</el-menu-item>
      </el-menu-item-group>
    </el-sub-menu>
    <el-sub-menu index="5">
      <template #title>
        <span>成绩管理</span>
      </template>
      <el-menu-item-group>
        <el-menu-item index="/score"><el-icon>
            <Menu />
          </el-icon>成绩列表</el-menu-item>
        <el-menu-item index="/addScore"><el-icon>
            <Goods />
          </el-icon>添加成绩</el-menu-item>
      </el-menu-item-group>
    </el-sub-menu>
  </el-menu>
```

```vue
        </el-aside>
        <el-container class="content">
          <Header />
          <div class="main">
            <router-view />
          </div>
          <Footer />
        </el-container>
      </el-container>
      <el-container v-else class="container">
        <router-view />
      </el-container>
    </div>
</template>

<script setup>
import Header from '@/components/Header.vue'
import { Avatar, FullScreen, Menu, UserFilled } from '@element-plus/icons-vue'
import { reactive } from 'vue'
import { useRouter } from 'vue-router'
import { localGet } from './utils'

const noMenu = ['/login']
const router = useRouter()
const state = reactive({
  showMenu: true,
  defaultOpen: ['1', '2', '3', '4', '5'],
  currentPath: "/",
})

router.beforeEach((to, from, next) => {
  if (to.path == '/login') {
    // 如果路径是 /login,则正常执行
    next()
  } else {
    // 如果不是 /login,判断是否有 token
    if (!localGet('token')) {
      // 如果没有,则跳至登录页面
      next({ path: '/login' })
    } else {
      // 否则继续执行
      next()
    }
  }
  state.showMenu = !noMenu.includes(to.path)
  state.currentPath = to.path
  document.title = pathMap[to.name]
  state.currentPath=to.path
  next()
})
</script>
```

```
<style scoped>
.layout {
  min-height: 100vh;
  background-color: #ffffff;
}

.container {
  height: 100vh;
}

.aside {
  width: 200px !important;
  background-color: #222832;
}

.head {
  display: flex;
  align-items: center;
  justify-content: center;
  height: 50px;
}

.head>div {
  display: flex;
  align-items: center;
}

.head img {
  width: 50px;
  height: 50px;
  margin-right: 10px;
}

.head span {
  font-size: 20px;
  color: #ffffff;
}

.line {
  border-top: 1px solid hsla(0, 0%, 100%, .05);
  border-bottom: 1px solid rgba(0, 0, 0, .2);
}

.content {
  display: flex;
  flex-direction: column;
  max-height: 100vh;
  overflow: hidden;
}

.main {
  height: calc(100vh - 100px);
```

```css
    overflow: auto;
    padding: 10px;
}
</style>

<style>
body {
    padding: 0;
    margin: 0;
    box-sizing: border-box;
}

.el-menu {
    border-right: none !important;
}

.el-submenu {
    border-top: 1px solid hsla(0, 0%, 100%, .05);
    border-bottom: 1px solid rgba(0, 0, 0, .2);
}

.el-submenu:first-child {
    border-top: none;
}

.el-submenu [class^="el-icon-"] {
    vertical-align: -1px !important;
}

a {
    color: #409eff;
    text-decoration: none;
}

.el-pagination {
    text-align: center;
    margin-top: 20px;
}

.el-popper__arrow {
    display: none;
}
</style>
```

(2) 在 school-admin-ui 的 src/components 文件夹下创建 Header.vue 文件，Header.vue 文件用来显示登录的管理员信息，文件代码如下：

```html
<template>
  <div class="header">
    <div class="left">
    </div>
    <div class="right">
      <div class="nickname">
        登录名：{{state.adminInfo && state.adminInfo.name}}
```

```
            <a style="cursor: pointer; margin-right: 10px" @click="logout()">
退出</a>
        </div>
    </div>
  </div>
</template>

<script setup>
import { onMounted, reactive } from 'vue'
import { useRouter } from 'vue-router'
import axios from '@/utils/axios'
import { localRemove } from '@/utils'

const router = useRouter()
const state = reactive({
  adminInfo: null // 管理员信息
})
// 初始化执行方法
  onMounted(() => {
  const pathname = window.location.hash.split('/')[1] || ''
  if (!['login'].includes(pathname)) {
    getAdminInfo()
  }
})
// 获取用户信息
const getAdminInfo = async () => {
  const adminInfo = await axios.get('/admin/info')
  state.adminInfo = adminInfo.data.data
}
 // 退出登录
const logout = () => {
  axios.delete('/admin/logout').then(() => {
    // 退出之后，将本地保存的 token 清理掉
    localRemove('token')
    // 回到登录页
    router.push({ path: '/login' })
  })
}

</script>

<style scoped>
  .header {
    height: 50px;
    border-bottom: 1px solid #e9e9e9;
    display: flex;
    justify-content: space-between;
    align-items: center;
    padding: 0 20px;
  }
</style>
```

（3）在 school-admin-ui 的 src/router 文件夹下找到 index.js 文件，在文件的

```
const routes = []
```
中添加前端系统的默认页面和登录页面的信息。
添加完成后的代码如下：

```
const routes = [
//定义系统默认页面路径，默认跳转到 home
  {
    path: '/',
    redirect: '/home'
  },
//定义 home 页面路径
  {
    path: '/home',
    name: 'home',
    component: () => import('../views/Home.vue')
  },
//定义登录页面路径
  {
    path: '/login',
    name: 'login',
    component: () => import('../views/Login.vue')
  },
]
```

(4) 创建 home 页面。在 school-admin-ui 的 src/views 文件夹下创建 Home.vue 文件，Home.vue 文件所定义的页面为登录进入系统的默认页面，编辑文件添加代码如下：

```
<template>
    <div class="home">
        <h1>欢迎来到教学信息管理系统</h1>
    </div>
</template>
```

(5) 创建登录页面。在 school-admin-ui 的 src/views 文件夹下中创建 Login.vue 文件，Login.vue 文件所定义的页面为登录页面，编辑文件添加代码如下：

```
<template>
  <div class="login-body">
    <div class="login-container">
      <div class="head">
        <div class="name">
          <div class="title">教学信息管理系统</div>
        </div>
      </div>
      <el-form label-position="top" :rules="state.rules" :model="state.ruleForm" ref="loginForm" class="login-form">
        <el-form-item label="账号" prop="username">
          <el-input type="text" v-model.trim="state.ruleForm.username" autocomplete="off"></el-input>
        </el-form-item>
        <el-form-item label="密码" prop="password">
```

```html
          <el-input type="password" v-model.trim="state.ruleForm.password"
autocomplete="off"></el-input>
        </el-form-item>
        <el-form-item>
          <div></div>
          <el-button style="width: 100%" type="primary" @click=
"submitForm">立即登录</el-button>
        </el-form-item>
      </el-form>
    </div>
  </div>
</template>
```

```javascript
<script setup>
import axios from '@/utils/axios';
import { reactive, ref } from 'vue';
import { localSet } from '../utils';
const loginForm = ref(null)
const state = reactive({
  ruleForm: {
    username: '',
    password: ''
  },
  checked: true,
  rules: {
    username: [
      { required: 'true', message: '账户不能为空', trigger: 'blur' }
    ],
    password: [
      { required: 'true', message: '密码不能为空', trigger: 'blur' }
    ]
  }
})
const submitForm = async () => {
  loginForm.value.validate((valid) => {
    if (valid) {
      axios.post('/admin/login', {
        name: state.ruleForm.username || '',
        password: state.ruleForm.password
      }).then(res => {
        if(res.data.responseCode==401){
          ElMessage.error('登录失败！')
        }else{
          localSet('token', res.data.data)
            window.location.href = '/'
        }

      })
    } else {
      console.log('error submit!!')
      return false;
    }
  })
```

```
}
const resetForm = () => {
  loginForm.value.resetFields();
}
</script>

<style scoped>
  .login-body {
    display: flex;
    justify-content: center;
    align-items: center;
    width: 100%;
    background-color: #fff;
  }
  .login-container {
    width: 420px;
    height: 500px;
    background-color: #fff;
    border-radius: 4px;
    box-shadow: 0px 21px 41px 0px rgba(0, 0, 0, 0.2);
  }
  .head {
    display: flex;
    justify-content: center;
    align-items: center;
    padding: 40px 0 20px 0;
  }
  .head img {
    width: 100px;
    height: 100px;
    margin-right: 20px;
  }
  .head .title {
    font-size: 28px;
    color: #1BAEAE;
    font-weight: bold;
  }
  .head .tips {
    font-size: 12px;
    color: #999;
  }
  .login-form {
    width: 70%;
    margin: 0 auto;
  }
  .login-form >>> .el-form--label-top .el-form-item__label {
    padding: 0;
  }
  .login-form >>> .el-form-item {
    margin-bottom: 0;
  }
</style>
```

经过以上步骤，管理员登录和登出所涉及的前端系统已经准备完毕，待后端系统代码开发完成之后会进行前后端联调，具体的页面显示效果到联调时展示。

### 2. 开发后端

后端系统在 10.3.2 节中进行了初始化，下面的步骤在后端系统 school-admin 中进行。

(1) 创建 admin 表的持久化类。

在 school-admin 的 src/main/java 目录下的 com.demo.school.bean 包中创建数据表 admin 的持久化类 Admin，代码如下：

```java
package com.demo.school.bean;

import java.util.Date;

public class Admin {
    //管理员id
    private Integer id;
    //管理员账户名
    private String name;
    //管理员密码
    private String password;
    //管理员token
    private String token;
    //管理员token过期时间
    private Date expiredTime;
    //getter 和 setter 方法省略
}
```

(2) 开发数据访问层。

在 school-admin 的 src/main/java 目录下的 com.demo.school.dao 包中创建数据访问接口 AdminDao，代码如下：

```java
package com.demo.school.dao;

import com.demo.school.bean.Admin;
import org.apache.ibatis.annotations.Mapper;
import org.springframework.stereotype.Repository;

@Repository
@Mapper
public interface AdminDao {
    //管理员登录，设置token和过期时间
    Admin login(Admin admin);
    //管理员登出，删除token和过期时间
    Integer logout(String token);
    //设置管理员的token
    Integer setToken(Admin admin);
    //通过token获取管理员的信息
    Admin getInfoByToken(String token);
}
```

在 school-admin 的 src/main/java/resources/mapper 文件夹下创建 AdminDao.xml 文件，并添加和 AdminDao 对应的 MyBatis 的数据库操作内容，代码如下：

```xml
<!DOCTYPE mapper
        PUBLIC "-//mybatis.org//DTD Mapper 3.0//EN"
        "https://mybatis.org/dtd/mybatis-3-mapper.dtd">
<mapper namespace="com.demo.school.dao.AdminDao">

    <select id="login" parameterType="Admin" resultType="Admin">
        SELECT * FROM admin WHERE name=#{name} AND password=md5(#{password});
    </select>
    <update id="setToken" parameterType="map" >
        UPDATE admin set token=#{token}, expired_time=#{expiredTime} WHERE id=#{id};
    </update>
    <update id="logout" parameterType="string" >
        UPDATE admin set token=null, expired_time=null WHERE token=#{token};
    </update>
    <select id="getInfoByToken" parameterType="string" resultType="Admin">
        SELECT * FROM admin WHERE token=#{token} AND expired_time>NOW();
    </select>
</mapper>
```

(3) 开发业务逻辑层。

在 school-admin 的 src/main/java 目录下的 com.demo.school.service 包中创建数据访问接口 AdminService，代码如下：

```java
package com.demo.school.service;

import com.demo.school.bean.Admin;

public interface AdminService {
    //管理员登录
    public String login(Admin admin);
    //管理员登出
    public Integer logout(String token);
    //获取管理员信息
    public Admin getInfoByToken(String token);
}
```

在 school-admin 的 src/main/java 目录下的 com.demo.school.service.impl 包中创建数据访问接口 AdminService 的实现类 AdminServiceImpl，代码如下：

```java
package com.demo.school.service.impl;

import com.demo.school.bean.Admin;
import com.demo.school.dao.AdminDao;
import com.demo.school.service.AdminService;
import org.springframework.beans.factory.annotation.Autowired;
import org.springframework.stereotype.Service;
import org.springframework.util.DigestUtils;

import java.util.Calendar;
import java.util.Date;
```

```java
import java.util.Random;

@Service
public class AdminServiceImpl implements AdminService {

    @Autowired
    private AdminDao adminDao;
    //管理员登录，验证账户密码
    //如果成功就创建token,设置token过期时间为1个小时
    //将token和过期时间保存到admin表对应的用户数据中并返回token给上一层
    //如果失败则返回null
    @Override
    public String login(Admin admin) {
        Admin adminResult = adminDao.login(admin);
        if (adminResult != null) {
            String token = this.makeToken();
            Date now = new Date();
            Calendar c = Calendar.getInstance();
            c.setTime(now);
            c.add(Calendar.MINUTE, 60);
            adminResult.setToken(token);
            adminResult.setExpiredTime(c.getTime());
            adminDao.setToken(adminResult);
            return token;
        } else {
            return null;
        }
    }
    //管理员登出
    @Override
    public Integer logout(String token){
        return adminDao.logout(token);
    }
    //通过请求token获取管理员信息
    @Override
    public Admin getInfoByToken(String token) {
        return adminDao.getInfoByToken(token);
    }
    //使用md5算法创建token
    private String makeToken() {
        String token=(System.currentTimeMillis() + new Random().nextInt(999999999)) + "";
        return DigestUtils.md5DigestAsHex(token.getBytes());
    }

}
```

(4) 开发请求处理层。

在 school-admin 的 src/main/java 目录下的 com.demo.school.controller 包中创建请求处理类 AdminController，代码如下：

```java
package com.demo.school.controller;

import com.demo.school.bean.Admin;
import com.demo.school.service.AdminService;
import com.demo.school.utils.Response;
import com.demo.school.utils.ResultGenerator;
import org.springframework.beans.factory.annotation.Autowired;
import org.springframework.http.HttpStatus;
import org.springframework.web.bind.annotation.*;

import javax.servlet.http.HttpServletRequest;
import javax.servlet.http.HttpServletResponse;

@RestController
@RequestMapping("/admin")
public class AdminController {
    @Autowired
    private AdminService adminService;
    //管理员登录,如果adminService返回的token为null,表示登录失败,返回401
    //如果adminService返回的token不为null,表示登录成功,返回token
    @PostMapping("/login")
    public Response login(@RequestBody Admin admin, HttpServletResponse resp){
        String adminToken= adminService.login(admin);
        if(adminToken!=null){
            return ResultGenerator.genSuccessResultData(adminToken);
        }else{
            return ResultGenerator.genFailResult(HttpStatus.UNAUTHORIZED.value(), "登录失败");
        }

    }
    //管理员登出,获取请求头的token,然后在数据库中将token删除
    @DeleteMapping("/logout")
    public Response logout(HttpServletRequest req, HttpServletResponse resp){
       String token = req.getHeader("Authorization");
       adminService.logout(token);
       return ResultGenerator.genSuccessResult();

    }
    //获取管理员信息，获取请求头的token,然后再以此查询管理员信息,token和密码不返回
    @GetMapping("/info")
    public Response getInfo(HttpServletRequest req, HttpServletResponse resp){
        String token = req.getHeader("Authorization");
        Admin admin=adminService.getInfoByToken(token);
        //只返回管理员名称,密码和token等敏感信息不返回,设置为null
        admin.setToken(null);
        admin.setExpiredTime(null);
```

```
        admin.setPassword(null);
        return ResultGenerator.genSuccessResultData(admin);
    }
}
```

**3. 测试功能**

（1） 按 7.1.2 小节中的方法启动后端系统，系统会在本地的 8080 开启 Web 服务。然后在前端系统 school-admin-ui 文件夹中打开命令行界面运行如下命令启动前端系统：

```
npm run dev
```

前端系统启动后，命令行窗口显示如图 10-16 所示。

（2） 从结果中可以看到前端系统的访问地址为 http://127.0.0.1:5173，打开浏览器输入登录页面访问地址 http://127.0.0.1:5173/login，登录页面如图 10-17 所示。

图 10-16　前端系统启动结果　　　　　图 10-17　教学信息管理系统登录页面

（3） 在账号和密码输入框中输入测试账户信息 admin 和 123456 后单击登录，页面跳转到主页面，如图 10-18 所示。

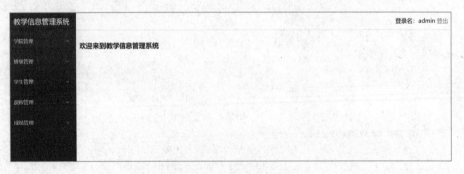

图 10-18　教学信息管理系统主页面

在主页面右上角显示登录账户的登录名为 admin。

（4） 单击主页面右上角的"登出"按钮，系统会跳转到登录页面。

## 10.5.2　管理员访问验证

在上一节中完成开发了管理员的登录功能，但是如果用户跳过登录页面直接访问 home 页面是可以跳过登录验证的，在实际系统中这是不可接受的。对于每一个进入后端系统的请求都应该检查该请求的客户端是不是已经登录了。在登录功能中，当前端登录成功后会将后端返回的 token 保存起来，后续的每次请求都可以将 token 放到请求中发送到后端，后端可以对请求进行拦截，获取请求的 token 并进行验证，验证成功就处理请求并返回结果，验证失败则返回代码 401 给前端，前端会自动跳转到登录页面。管理员访问验证流程如图 10-19 所示。

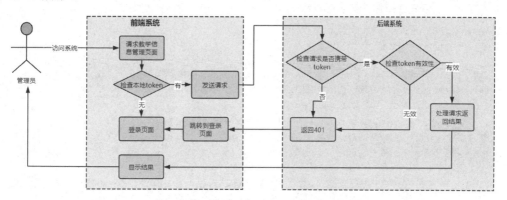

图 10-19　管理员访问验证流程

在本书的第 9 章中已经介绍了拦截器及其使用，下面就使用拦截器来实现管理员的访问验证功能。

**1．创建访问验证拦截器**

在 school-admin 的 src/main/java 目录下的 com.demo.school.interceptor 包中创建拦截器 AuthenticationInterceptor，代码如下：

```java
package com.demo.school.interceptor;

import com.demo.school.service.AdminService;
import com.demo.school.utils.Response;
import com.demo.school.utils.ResultGenerator;
import com.fasterxml.jackson.databind.ObjectMapper;
import org.apache.commons.lang3.StringUtils;
import org.springframework.beans.factory.annotation.Autowired;
import org.springframework.stereotype.Component;
import org.springframework.web.method.HandlerMethod;
import org.springframework.web.servlet.HandlerInterceptor;

import javax.servlet.http.HttpServletRequest;
import javax.servlet.http.HttpServletResponse;
import java.io.PrintWriter;

@Component
public class AuthenticationInterceptor implements HandlerInterceptor {

    @Autowired
    private AdminService adminService;

    @Override
    public boolean preHandle(HttpServletRequest request, HttpServletResponse
```

```java
response, Object handler)
        throws Exception {
    if (!(handler instanceof HandlerMethod)) {
        return true;
    }
    String token = request.getHeader("Authorization");
     //token 为空返回 401,认证失败
    if (StringUtils.isEmpty(token)) {
        ObjectMapper mapper = new ObjectMapper();
        Response result = ResultGenerator.genFailResult(ResultGenerator.RESULT_CODE_UNAUTHORIZED_ERROR, "token is null or empty");
        PrintWriter out = response.getWriter();
        out.append(mapper.writeValueAsString(result));
        return false;
    }
    //通过 token 获取管理员信息,如果返回值为 null,说明 token 是错误或者是过期的
    else if (null==this.adminService.getInfoByToken(token)) {
        //token 错误或者过期返回 401,认证失败
        ObjectMapper mapper = new ObjectMapper();
        Response result = ResultGenerator.genFailResult(ResultGenerator.RESULT_CODE_UNAUTHORIZED_ERROR, "invalid token");
        PrintWriter out = response.getWriter();
        out.append(mapper.writeValueAsString(result));
        return false;
    } else {
        //认证成功
        return true;
    }

}
@Override
public void afterCompletion(HttpServletRequest request, HttpServletResponse
  response, Object handler, Exception ex) throws Exception {

}
}
```

### 2. 配置访问验证拦截器

在项目的 src/main/java 目录下的 com.demo.school.configuration 包中,编辑 SchoolConfiguration 类,在其中添加拦截器的配置,代码如下:

```java
// addPathPatterns 是拦截所有路径,excludePathPatterns 是排除需要拦截的路径
// 添加拦截器,登录请求是不可能有 token 的,所以要排除登录请求
@Override
public void addInterceptors(InterceptorRegistry registry) {
    registry.addInterceptor(authenticationInterceptor)
        .addPathPatterns("/**")
        .excludePathPatterns("/admin/login");
}
```

### 3. 测试功能

重新启动前端系统和后端系统。在未登录的情况下访问系统的 home 页面会自动跳转到登录页面，在登录成功的情况下访问 home 页面会正确进入页面。

## 10.6 教学信息管理模块

教学信息管理模块包括学院管理、班级管理、学生管理、课程管理和成绩管理五个模块。

### 10.6.1 学院管理

学院管理功能包含学院信息的列表查询、添加、修改和删除。

#### 1. 开发前端

(1) 创建学院列表页面。

在 school-admin-ui 的 src/views 文件夹下创建 College.vue 文件，代码如下：

```vue
<template>
  <el-card class="good-container">
   <template #header>
    <div class="header">
      <el-button type="primary" :icon="Plus" @click="handleAdd()">添加学院</el-button>
    </div>
   </template>
   <el-table
     :data="state.tableData"
     tooltip-effect="dark"
     style="width: 100%"
   >
     <el-table-column
       prop="id"
       label="编号"
     >
     </el-table-column>
     <el-table-column
       prop="name"
       label="学院名称"
     >
     </el-table-column>

     <el-table-column
       label="操作"
       width="100"
     >
       <template #default="scope">
```

```html
            <a style="cursor: pointer; margin-right: 10px" @click="handleEdit
(scope.row.id)">修改</a>
          <el-popconfirm
             title="确定删除吗？"
             confirmButtonText='确定'
             cancelButtonText='取消'
             @confirm="handleDelete(scope.row.id)"
          >
             <template #reference>
                <a style="cursor: pointer">删除</a>
             </template>
          </el-popconfirm>
        </template>
      </el-table-column>
    </el-table>
    <!--总数超过一页，则展示分页器-->
    <el-pagination
      v-model:current-page="state.currentPage"
      v-model:page-size="state.pageSize"
      :page-sizes="[10, 20, 50, 100]"
      :small="small"
      :disabled="disabled"
      :background="background"
      layout="sizes, prev, pager, next"
      :total="state.total"
      @size-change="changeSize"
      @current-change="changePage"
    />
  </el-card>
</template>
```

```js
<script setup>
import axios from '@/utils/axios'
import { Plus } from '@element-plus/icons-vue'
import { ElMessage } from 'element-plus'
import { getCurrentInstance, onMounted, onUnmounted, reactive } from 'vue'
import { useRouter } from 'vue-router'
import { localGet, localRemove, localSet } from '../utils'

const app = getCurrentInstance()
const router = useRouter()
const state = reactive({
  loading: false,
  tableData: [], // 数据列表
  total: 0, // 总条数
  currentPage: 1, // 当前页
  pageSize: 10 // 分页大小
})
onMounted(() => {
  getCollegeList()
  state.currentPage=localGet("currentPage")===null?1:localGet("currentPage")
```

```js
    state.pageSize=localGet("pageSize")===null?10:localGet("pageSize")
})
onUnmounted(() => {
  localRemove("currentPage");
  localRemove("pageSize");
});
const getCollegeList = () => {
  state.loading = true
  axios.get('/college/list', {
    params: {
      pageNumber: state.currentPage,
      pageSize: state.pageSize
    }
  }).then(res => {
    state.tableData = res.data.data.list
    state.total = res.data.data.total
    state.currentPage = res.data.data.currentPage
  })
}
const handleAdd = () => {
  router.push({ path: '/addCollege' })
}
const handleEdit = (id) => {
  router.push({ path: '/addCollege', query: { id } })
}
const handleDelete = (id) => {
    axios.delete(`/college/${id}/delete`, {
  }).then(() => {
    ElMessage.success('删除成功')
    getCollegeList()
  })
}
const changePage = (val) => {
  state.currentPage = val
  localSet("currentPage",val)
  getCollegeList()
}
const changeSize = (val) => {
  state.pageSize = val
  localSet("pageSize",val)
  getCollegeList()
}

</script>

<style scoped>
  .good-container {
    min-height: 100%;
  }
  .el-card.is-always-shadow {
    min-height: 100%!important;
```

    }
</style>
```

(2) 创建添加和编辑学院页面。

在 school-admin-ui 的 src/views 文件夹下创建 AddCollege.vue 文件，代码如下：

```
<template>
  <div class="add">
    <el-card class="add-container">
        <el-form  :model="state.collegeForm"   :rules="state.rules" ref="collegeRef" label-width="100px" class="collegeForm">
       <el-form-item label="学院" prop="college">
         <el-input style="width: 300px" v-model="state.collegeForm.name" placeholder="请输学院名称"></el-input>
       </el-form-item>
       <el-form-item>
         <el-button type="primary" @click="submitAdd()">{{ state.id ? '立即修改' : '立即添加' }}</el-button>
       </el-form-item>
     </el-form>
   </el-card>
  </div>
</template>

<script setup>
import axios from '@/utils/axios';
import { ElMessage } from 'element-plus';
import { onMounted, reactive, ref } from 'vue';
import { useRoute } from 'vue-router';

const collegeRef = ref(null)
const route = useRoute()
const { id } = route.query
const state = reactive({
//通过id来区分是添加还是修改，访问页面时传递id为修改，否则为添加
  id: id,
  collegeForm: {
    name: ''
  },
  rules: {
    name: [
      { required: 'true', message: '请填写学院名称', trigger: ['change'] }
    ]
  },
})
onMounted(() => {
  if (id) {
    axios.get(`/college/${id}`).then(res => {
      state.collegeForm = {
        name: res.data.data.name,
      }
    })
  }
})
```
```

```
const submitAdd = () => {
  collegeRef.value.validate((vaild) => {
    if (vaild) {
      let params = {
        name: state.collegeForm.name
      }
      console.log('params', params)
      if (id) {
        params.id = id
        axios.put(`/college/update`, params).then(() => {
          ElMessage.success('修改成功')
        })
      } else {
        axios.post(`/college/add`, params).then(() => {
          ElMessage.success('添加成功')
        })
      }
    }
  })
}
</script>

<style scoped>
.add {
  display: flex;
}

.add-container {
  flex: 1;
  height: 100%;
}

.avatar-uploader {
  width: 100px;
  height: 100px;
  color: #ddd;
  font-size: 30px;
}

.avatar-uploader-icon {
  display: block;
  width: 100%;
  height: 100%;
  border: 1px solid #e9e9e9;
  padding: 32px 17px;
}
</style>
```

(3) 注册学院列表与添加页面。

在 school-admin-ui 的 src/router 文件夹下找到 index.js 文件，在文件的

```
const routes = []
```

中添加如下学院列表和添加学院两个页面：

```
    {
      path: '/college',
      name: 'college',
      component: () => import('../views/College.vue')
    },
    {
      path: '/addCollege',
      name: 'addCollege',
      component: () => import('../views/AddCollege.vue')
    },
```

2. 开发后端

(1) 创建 college 表的持久化类。

在 school-admin 的 src/main/java 目录下的 com.demo.school.bean 包中创建数据表 college 的持久化类 College，代码如下：

```java
package com.demo.school.bean;

import java.io.Serializable;

public class College implements Serializable {
    //学院 id
    private Integer id;
    //学院名称
    private String name;
    //getter 和 setter 方法省略
}
```

(2) 开发数据访问层。

在 school-admin 的 src/main/java 目录下的 com.demo.school.dao 包中创建数据访问接口 CollegeDao，代码如下：

```java
package com.demo.school.dao;

import com.demo.school.bean.College;
import com.demo.school.utils.PageQuery;
import org.apache.ibatis.annotations.Mapper;
import org.springframework.stereotype.Repository;

import java.util.List;
/*
学院信息数据访问接口
 */
@Mapper
@Repository
public interface CollegeDao {
    //分页查询学院列表
    List<College> list(PageQuery pageQuery);
    //查询所有学院列表
    List<College> getAll();
```

```
    //查询所有学院总数
    Integer getTotal();
    //查询学院信息
    College getById(Integer id);
    //添加学院信息
    Integer add(College college);
    //修改学院信息
    Integer update(College college);

    //删除学院信息
    Integer delete(Integer id);
}
```

在 school-admin 的 src/main/java/resources/mapper 文件夹下创建 CollegeDao.xml 文件，并添加和 CollegeDao 接口对应的 MyBatis 的数据库操作内容，代码如下：

```xml
<!DOCTYPE mapper
    PUBLIC "-//mybatis.org//DTD Mapper 3.0//EN"
    "https://mybatis.org/dtd/mybatis-3-mapper.dtd">
<mapper namespace="com.demo.school.dao.CollegeDao">
    <select id="getById" parameterType="int" resultType="College">
        SELECT * FROM college WHERE id=#{id};
    </select>
    <select id="list" parameterType="PageQuery" resultType="College">
        SELECT * FROM college ORDER
        BY id ASC LIMIT #{start},#{pageSize};
    </select>
    <select id="getAll" resultType="College">
        SELECT * FROM college ORDER BY id ASC;
    </select>
    <select id="getTotal" resultType="int">
        SELECT COUNT(*) FROM college;
    </select>
    <insert id="add" parameterType="College">
        INSERT INTO college(name) VALUES(#{name});
    </insert>
    <update id="update" parameterType="College">
        UPDATE college SET name=#{name} WHERE ID=${id};
    </update>
    <delete id="delete" parameterType="int">
        DELETE FROM college WHERE id = #{id};
    </delete>
</mapper>
```

（3）开发业务逻辑层。

在 school-admin 的 src/main/java 目录下的 com.demo.school.service 包中创建数据访问接口 CollegeService，代码如下：

```java
package com.demo.school.service;

import com.demo.school.bean.College;
import com.demo.school.utils.PageQuery;
```

```java
import com.demo.school.utils.PageResponse;

import java.util.List;

/*
学院信息 Service 接口
 */
public interface CollegeService {
    //分页查询学院列表
    PageResponse list(PageQuery pageQuery);
    //查询学院信息
    College getById(int id);
    //添加学院信息
    Integer add(College college);
    //修改学院信息
    Integer update(College college);
    //删除学院信息
    Integer delete(Integer id);
    //查询所有学院信息列表
    List<College> getAll();
}
```

在 school-admin 的 src/main/java 目录下的 com.demo.school.service.impl 包中创建数据访问接口 CollegeService 的实现类 CollegeServiceImpl，代码如下：

```java
package com.demo.school.service.impl;

import com.demo.school.bean.College;
import com.demo.school.dao.CollegeDao;
import com.demo.school.service.CollegeService;
import com.demo.school.utils.PageQuery;
import com.demo.school.utils.PageResponse;
import org.springframework.beans.factory.annotation.Autowired;
import org.springframework.stereotype.Service;

import java.util.List;

/*
学院信息 Service 实现类
 */
@Service
public class CollegeServiceImpl implements CollegeService {

    //注入 CollegeDao 对象
    @Autowired
    private CollegeDao collegeDao;

    //分页查询学院列表
    @Override
    public PageResponse list(PageQuery pageQuery) {
        //根据分页查询封装对象，查询当前页学院数据列表
        List<College> list = this.collegeDao.list(pageQuery);
```

```java
    //查询学院总数
    int total = this.collegeDao.getTotal();
    //构建分页查询结果
    PageResponse pageResult = new PageResponse(
            list, total, pageQuery.getPageSize(), pageQuery.getPageNumber());
    return pageResult;
}

//查询学院信息
@Override
public College getById(int id) {
    return collegeDao.getById(id);
}

//添加学院信息
@Override
public Integer add(College college) {
    return collegeDao.add(college);
}

//修改学院信息
@Override
public Integer update(College college) {
    return collegeDao.update(college);
}

//删除学院信息
@Override
public Integer delete(Integer id) {
    return collegeDao.delete(id);
}

//查询所有学院信息列表
@Override
public List<College> getAll() {
    return this.collegeDao.getAll();
}
}
```

(4) 开发请求处理层。

在 school-admin 的 src/main/java 目录下的 com.demo.school.controller 包中创建请求处理类 CollegeController，其中的分页查询和响应结果使用 10.4 节中已经封装好的代码，代码如下：

```java
package com.demo.school.controller;

import com.demo.school.bean.College;
import com.demo.school.service.CollegeService;
import com.demo.school.utils.PageQuery;
import com.demo.school.utils.Response;
import com.demo.school.utils.ResultGenerator;
import org.springframework.beans.factory.annotation.Autowired;
import org.springframework.stereotype.Controller;
```

```java
import org.springframework.web.bind.annotation.*;
/*
学院信息Controller
 */
@Controller
@RequestMapping("/college")
public class CollegeController {
    // 注入CollegeService对象
    @Autowired
    private CollegeService collegeService;

    //分页查询学院列表
    @RequestMapping("list")
    @ResponseBody
    public Response getCollegeList(@RequestParam(required = false) Integer pageNumber,@RequestParam(required = false) Integer pageSize) {
        //验证分页参数，页码pageNumber和每页数据数pageSize必须提供
        //pageNumber最小为1，pageSize最小为10
        if (pageNumber == null || pageNumber < 1
                || pageSize == null || pageSize < 10) {
            //分页参数错误返回code为BadRequest 400
            return ResultGenerator.genFailResult(
            ResultGenerator.RESULT_CODE_BAD_REQUEST_ERROR, "分页参数异常！");
        }
        //创建分页查询参数
        PageQuery pageQuery = new PageQuery(pageNumber, pageSize);
        //调用分页查询并发挥成功结果
        return ResultGenerator.genSuccessResultData(
                collegeService.list(pageQuery));
    }

    //查询所有学院信息列表
    @RequestMapping("all")
    @ResponseBody
    public Response getAll() {
        return ResultGenerator.genSuccessResultData(
                this.collegeService.getAll());
    }

    //查询学院信息
    @RequestMapping(path = "/{id}", method = {RequestMethod.GET})
    @ResponseBody
    public Response getById(@PathVariable Integer id) {
        return ResultGenerator.genSuccessResultData(
                this.collegeService.getById(id));
    }

    //添加学院信息
    @RequestMapping(path = "/add", method = {RequestMethod.POST})
    @ResponseBody
    public Response addCollege(@RequestBody College college) {
        return ResultGenerator.genSuccessResultData(
```

```
        collegeService.add(college));
}

//修改学院信息
@RequestMapping(path = "/update", method = {RequestMethod.PUT})
@ResponseBody
public Response updateCollege(@RequestBody College college) {
    return ResultGenerator.genSuccessResultData(
        collegeService.update(college));
}

//删除学院信息
@RequestMapping(path = "/{id}/delete", method = {RequestMethod.DELETE})
@ResponseBody
public Response deleteCollege(@PathVariable Integer id) {
    return ResultGenerator.genSuccessResultData(
        collegeService.delete(id));
}
}
```

**3. 测试功能**

(1) 启动前端系统和后端系统并访问前端系统，登录系统。

(2) 在页面左侧访问"学院管理"→"学院列表"查看学院列表，如图 10-20 所示，当前数据库中只有 5 条学院信息，页面默认每页显示 10 条数据，所以只有 1 页数据。

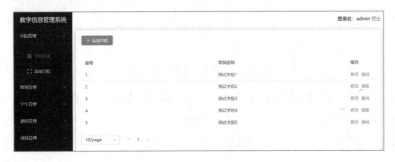

图 10-20　查看学院列表页面 1

(3) 单击系统左侧的学院管理列表中的"添加学院"命令或者单击学院列表数据上面的"添加学院"按钮，进入添加学院页面，输入学院名称并单击"立即添加"按钮，即可添加学院数据，如图 10-21 所示。

图 10-21　添加学院页面

(4) 按照步骤(3)的操作多添加几个学院,学院总数超过 10 之后再返回学院列表页面,页面可以显示分页效果,如图 10-22 所示,页面中显示有 2 页。

| 编号 | 学院名称 | 操作 |
| --- | --- | --- |
| 1 | 测试学院1 | 修改 删除 |
| 2 | 测试学院2 | 修改 删除 |
| 3 | 测试学院3 | 修改 删除 |
| 4 | 测试学院4 | 修改 删除 |
| 5 | 测试学院5 | 修改 删除 |
| 6 | 测试学院6 | 修改 删除 |
| 7 | 测试学院7 | 修改 删除 |
| 8 | 测试学院8 | 修改 删除 |
| 9 | 测试学院9 | 修改 删除 |
| 10 | 测试学院10 | 修改 删除 |

10/page  < 1 2 >

图 10-22 查看学院列表页面 2

(5) 单击页面下方的页码表中的"2"进入第二页数据,如图 10-23 所示。

| 编号 | 学院名称 | 操作 |
| --- | --- | --- |
| 11 | 测试学院11 | 修改 删除 |
| 12 | 测试学院12-1 | 修改 删除 |

10/page  < 1 2 >

图 10-23 查看学院列表页面 3

(6) 在页面左下方设置每页显示数量的下拉列表中选择 20,设置每页显示 20 条数据,如图 10-24 所示。

| 编号 | 学院名称 | 操作 |
| --- | --- | --- |
| 1 | 测试学院1 | 修改 删除 |
| 2 | 测试学院2 | 修改 删除 |
| 3 | 测试学院3 | 修改 删除 |
| 4 | 测试学院4 | 修改 删除 |
| 5 | 测试学院5 | 修改 删除 |
| 6 | 测试学院6 | 修改 删除 |
| 7 | 测试学院7 | 修改 删除 |
| 8 | 测试学院8 | 修改 删除 |
| 9 | 测试学院9 | 修改 删除 |
| 10 | 测试学院10 | 修改 删除 |
| 11 | 测试学院11 | 修改 删除 |
| 12 | 测试学院12 | 修改 删除 |

20/page  < 1 >

图 10-24 查看学院列表页面 4

（7）在学院列表中选择一项然后单击右侧的"修改"按钮，进入修改学院页面，如图 10-25 所示。

（8）在修改页面中修改学院名称并单击"立即修改"按钮，即可修改学院信息，如图 10-26 所示。

图 10-25　修改学院页面 1

图 10-26　修改学院页面 2

（9）在学院列表中选择一项然后单击右侧的"删除"按钮，提示是否确定删除，如图 10-27 所示。

图 10-27　删除学院 1

（10）单击"确定"按钮，删除选中的学院信息，如图 10-28 所示。

图 10-28　删除学院 2

## 10.6.2 班级管理

班级管理功能包含班级信息的列表查询、添加、修改和删除。

### 1. 开发前端

(1) 创建班级列表页面。

在 school-admin-ui 的 src/views 文件夹下创建 Class.vue 文件，代码如下：

```
<template>
  <el-card class="good-container">
    <template #header>
      <div class="header">
        <el-button type="primary" :icon="Plus" @click="handleAdd">添加班级</el-button>
      </div>
    </template>
    <el-table :data="state.tableData" tooltip-effect="dark" style="width: 100%">
      <el-table-column prop="id" label="编号">
      </el-table-column>
      <el-table-column prop="name" label="班级名称">
      </el-table-column>
      <el-table-column prop="college.name" label="学院">
      </el-table-column>

      <el-table-column label="操作" width="100">
        <template #default="scope">
          <a style="cursor: pointer; margin-right: 10px" @click="handleEdit(scope.row.id)">修改</a>
          <el-popconfirm
            title="确定删除吗？"
            confirmButtonText='确定'
            cancelButtonText='取消'
            @confirm="handleDelete(scope.row.id)"
          >
            <template #reference>
              <a style="cursor: pointer">删除</a>
            </template>
          </el-popconfirm>
        </template>
      </el-table-column>
    </el-table>
    <!--总数超过一页,则展示分页器-->
    <el-pagination v-model:current-page="state.currentPage" v-model:page-size="state.pageSize" :page-sizes="[10, 20, 50, 100]"
      :small="small" :disabled="disabled" :background="background"
      layout="sizes, prev, pager, next" :total="state.total"
      @size-change="changeSize" @current-change="changePage" />
  </el-card>
</template>
```

```
<script setup>
import axios from '@/utils/axios'
import { Plus } from '@element-plus/icons-vue'
import { ElMessage } from 'element-plus'
import { getCurrentInstance, onMounted, onUnmounted, reactive } from 'vue'
import { useRouter } from 'vue-router'
import { localGet, localRemove, localSet } from '../utils'

const app = getCurrentInstance()
const { goTop } = app.appContext.config.globalProperties
const router = useRouter()
const state = reactive({
  loading: false,
  tableData: [],      //数据列表
  total: 0,           //总条数
  currentPage: 1,     //当前页
  pageSize: 10        //分页大小
})

onMounted(() => {
  state.currentPage=localGet("currentPage")===null?1:localGet("currentPage")
  state.pageSize=localGet("pageSize")===null?10:localGet("pageSize")
  getClassList()
})
onUnmounted(() => {
  localRemove("currentPage");
  localRemove("pageSize");
});
const getClassList = () => {
  state.loading = true
  axios.get('/class/list', {
    params: {
      pageNumber: state.currentPage,
      pageSize: state.pageSize
    }
  }).then(res => {
    state.tableData = res.data.data.list
    state.total = res.data.data.total
    state.currentPage = res.data.data.currentPage
    state.loading = false
    goTop && goTop()
  })
}
const handleAdd = () => {
  router.push({ path: '/addClass' })
}
const handleEdit = (id) => {
  router.push({ path: '/addClass', query: { id } })
}
const handleDelete = (id) => {
  axios.delete(`/class/${id}/delete`, {
  }).then(() => {
    ElMessage.success('删除成功')
```

```
      getClassList()
  })
}
const changePage = (val) => {
  state.currentPage = val
  localSet("currentPage",val)
  getClassList()
}
const changeSize = (val) => {
  state.pageSize = val
  localSet("pageSize",val)
  getClassList()
}
</script>

<style scoped>
.good-container {
  min-height: 100%;
}

.el-card.is-always-shadow {
  min-height: 100% !important;
}
</style>
```

(2) 创建添加班级页面。

在 school-admin-ui 的 src/views 文件夹下创建 AddClass.vue 文件,代码如下:

```
<template>
  <div class="add">
    <el-card class="add-container">
         <el-form    :model="state.classForm"    :rules="state.rules" ref="classRef" label-width="100px" class="classForm">
           <el-form-item label="班级" prop="class">
             <el-input style="width: 300px" v-model="state.classForm.name" placeholder="请输班级名称"></el-input>
           </el-form-item>
           <!-- 学院从列表选择 -->
           <el-form-item label="学院" prop="class">
             <el-select v-model="state.collegeValue" value-key="name" class="m-2"style="width: 300px;" placeholder="学院" size="large">
               <el-option v-for="item in state.collegeList":key= "item.id":label="item.name" :value="item" />
             </el-select>
           </el-form-item>
           <el-form-item>
             <el-button type="primary" @click="submitAdd()">{{ state.id ? '立即修改' : '立即添加' }}</el-button>
           </el-form-item>
         </el-form>
    </el-card>
  </div>
```

```
</template>

<script setup>
import axios from '@/utils/axios';
import { ElMessage } from 'element-plus';
import { onMounted, reactive, ref } from 'vue';
import { useRoute } from 'vue-router';

const classRef = ref(null)
const route = useRoute()
const { id } = route.query
const state = reactive({
  id: id,
  classForm: {
    name: '',
    collegeId: 0,
  },
  collegeList: [],
  collegeValue: {},
  rules: {
    name: [
      { required: 'true', message: '请填写班级名称', trigger: ['change'] }
    ]
  },
})
onMounted(() => {
//获取所有学院列表并赋值给单选框
  axios.get(`/college/all`).then(res => {
    state.collegeList = res.data.data
  })
  if (id) {
    axios.get(`/class/${id}`).then(res => {

      state.classForm = {
        name: res.data.data.name,
        collegeId: res.data.data.college.id,
      }
      axios.get(`/college/` + state.classForm.collegeId).then(res => {
        state.collegeValue = res.data.data;
      })

    })
  }

})
const submitAdd = () => {
  classRef.value.validate((vaild) => {
    if (vaild) {
      let params = {
        name: state.classForm.name,
        college:{
          id:state.collegeValue.id,
        },
```

```
      }
      console.log('params', params)
      if (id) {
        params.id = id
        axios.put(`/class/update`, params).then(() => {
          ElMessage.success('修改成功')
        })
      } else {
        axios.post(`/class/add`, params).then(() => {
          ElMessage.success('添加成功')
        })
      }
    }
  })
}
</script>

<style scoped>
.add {
  display: flex;
}

.add-container {
  flex: 1;
  height: 100%;
}

.avatar-uploader {
  width: 100px;
  height: 100px;
  color: #ddd;
  font-size: 30px;
}

.avatar-uploader-icon {
  display: block;
  width: 100%;
  height: 100%;
  border: 1px solid #e9e9e9;
  padding: 32px 17px;
}
</style>
```

（3）注册班级列表与添加页面。

在 school-admin-ui 的 src 文件夹的子文件夹 router 中找到 index.js 文件，在文件的

```
const routes = []
```

中添加如下班级列表和添加班级两个页面：

```
  {
    path: '/class',
    name: 'class',
    component: () => import('../views/Class.vue')
```

```
  },
  {
    path: '/addClass',
    name: 'addClass',
    component: () => import('../views/AddClass.vue')
  },
```

2. 开发后端

(1) 创建 class 表的持久化类。

在 school-admin 的 src/main/java 目录下的 com.demo.school.bean 包中创建数据表 class 的持久化类 Class，代码如下：

```
package com.demo.school.bean;

import java.io.Serializable;

public class Class implements Serializable {
    //班级 id
    private Integer id;
    //班级名称
    private String name;
    //关联的学院
//getter 和 setter 方法省略
}
```

(2) 开发数据访问层。

在 school-admin 的 src/main/java 目录下的 com.demo.school.dao 包中创建数据访问接口 ClassDao，代码如下：

```
package com.demo.school.dao;

import com.demo.school.bean.Class;
import com.demo.school.utils.PageQuery;
import org.apache.ibatis.annotations.Mapper;
import org.springframework.stereotype.Repository;

import java.util.List;

/*
班级信息数据访问接口
*/
@Repository
@Mapper
public interface ClassDao {
    //分页查询班级列表
    List<Class> list(PageQuery pageQuery);

    //查询所有班级列表
    List<Class> getAll(Integer collegeId);
```

```
    //查询所有班级总数
    Integer getTotal();

    //查询班级信息
    Class getById(Integer id);

    //添加班级信息
    Integer add(Class clazz);

    //修改班级信息
    Integer update(Class clazz);

    //删除班级信息
    Integer delete(Integer id);
}
```

在 school-admin 的 src/main/java/resources/mapper 文件夹下创建 ClassDao.xml 文件,并添加和 ClassDao 接口对应的 MyBatis 的数据库操作内容,班级 class 和学院 college 是一对一的关联关系,所以需要定义 resultMap 实现关联,代码如下:

```xml
<!DOCTYPE mapper
        PUBLIC "-//mybatis.org//DTD Mapper 3.0//EN"
        "https://mybatis.org/dtd/mybatis-3-mapper.dtd">
<mapper namespace="com.demo.school.dao.ClassDao">
<!--class 中学院是一对一的关联关系-->
    <resultMap id="classResultMap" type="Class">
        <id property="id" column="id"/>
        <result property="name" column="name"/>
        <association property="college" javaType="College">
            <id property="id" column="college_id"/>
            <result property="name" column="college_name"/>
        </association>
    </resultMap>
    <select id="getById" parameterType="int" resultMap="classResultMap">
        SELECT a.*,b.name as college_name FROM
        class a inner join college b
        on a.college_id=b.id WHERE a.id=#{id};
    </select>
    <select id="list" parameterType="PageQuery" resultMap="classResultMap">
        SELECT a.*,b.name as college_name FROM
        class a inner join college b
        on a.college_id=b.id
        ORDER BY id asc limit #{start},#{pageSize};
    </select>
    <select id="getAll" resultType="Class">
        SELECT * FROM class WHERE 1=1
        <if test="collegeId!=null">AND college_id=#{collegeId}</if>
        ORDER BY id ASC;
    </select>
    <select id="getTotal" resultType="int">
```

```xml
        SELECT COUNT(*) FROM class;
    </select>
    <insert id="add" parameterType="Class">
        INSERT INTO class(name,college_id)
        VALUES(#{name},#{college.id});
    </insert>
    <update id="update" parameterType="Class">
        UPDATE class SET
        name=#{name},college_id=#{college.id}
        WHERE ID=${id};
    </update>
    <delete id="delete" parameterType="int">
        DELETE FROM class WHERE ID = #{id};
    </delete>
</mapper>
```

(3) 开发业务逻辑层。

在 school-admin 的 src/main/java 目录下的 com.demo.school.service 包中创建数据访问接口 ClassService，代码如下：

```java
package com.demo.school.service;

import com.demo.school.utils.PageQuery;
import com.demo.school.utils.PageResponse;
import com.demo.school.bean.Class;

import java.util.List;

/*
班级信息 Service 接口
*/
public interface ClassService {
    //分页查询班级列表
    PageResponse list(PageQuery pageQuery);

    //查询班级信息
    Class getById(int id);

    //添加班级信息
    Integer add(Class theClass);

    //修改班级信息
    Integer update(Class theClass);

    //删除班级信息
    Integer delete(Integer id);

    //查询所有班级信息列表
    List<Class> getAll(Integer collegeId);
}
```

在 school-admin 的 src/main/java 目录下的 com.demo.school.service.impl 包中创建数据访

问接口 ClassService 的实现类 ClassServiceImpl，代码如下：

```java
package com.demo.school.service.impl;

import com.demo.school.bean.Class;
import com.demo.school.dao.ClassDao;
import com.demo.school.service.ClassService;
import com.demo.school.utils.PageQuery;
import com.demo.school.utils.PageResponse;
import org.springframework.beans.factory.annotation.Autowired;
import org.springframework.stereotype.Service;

import java.util.List;
/*
班级信息 Service 实现类
 */
@Service
public class ClassServiceImpl implements ClassService {

    // 注入 ClassDao 对象
    @Autowired
    private ClassDao classDao;

    // 分页查询班级列表
    @Override
    public PageResponse list(PageQuery pageQuery) {
        List<Class> list = this.classDao.list(pageQuery);
        int total = this.classDao.getTotal();
        PageResponse pageResult = new PageResponse(
                list, total, pageQuery.getPageSize(), pageQuery.getPageNumber());
        return pageResult;
    }

    //查询班级信息
    @Override
    public Class getById(int id) {
        return classDao.getById(id);
    }

    //添加班级信息
    @Override
    public Integer add(Class theClass) {
        return classDao.add(theClass);
    }

    //修改班级信息
    @Override
    public Integer update(Class theClass) {
        return classDao.update(theClass);
    }
```

```
    //删除班级信息
    @Override
    public Integer delete(Integer id) {
        return classDao.delete(id);
    }

    //查询所有班级信息列表
    @Override
    public List<Class> getAll(Integer collegeId) {
        return this.classDao.getAll(collegeId);
    }
}
```

(4) 开发请求处理层。

在 school-admin 的 src/main/java 目录下的 com.demo.school.controller 包中创建请求处理类 ClassController，其中的分页查询和响应结果使用 10.4 节中已经封装好的代码，代码如下：

```
package com.demo.school.controller;

import com.demo.school.service.ClassService;
import com.demo.school.bean.Class;
import com.demo.school.utils.PageQuery;
import com.demo.school.utils.Response;
import com.demo.school.utils.ResultGenerator;
import org.springframework.beans.factory.annotation.Autowired;
import org.springframework.web.bind.annotation.*;

import java.util.HashMap;
import java.util.Map;

/*
班级信息 Controller
 */
@RestController
@RequestMapping("/class")
@CrossOrigin
public class ClassController {
    //注入 ClassService 对象
    @Autowired
    private ClassService classService;

    //分页查询班级列表
    @RequestMapping("list")
    @ResponseBody
    public Response getClassList(@RequestParam(required = false) Integer pageNumber,
                 @RequestParam(required = false) Integer pageSize) {
        //验证分页参数，页码 pageNumber 和每页数据数 pageSize 必须提供
        //pageNumber 最小为 1，pageSize 最小为 10
        if (pageNumber == null || pageNumber < 1
                || pageSize == null || pageSize < 10) {
```

```java
        //请求中如果分页参数错误则返回BadRequest
        return ResultGenerator.genFailResult(ResultGenerator.RESULT_CODE_BAD_REQUEST_ERROR, "分页参数异常！");
    }
    //创建分页查询参数
    PageQuery pageQuery = new PageQuery(pageNumber, pageSize);
    //调用分页查询并发挥成功结果
    return ResultGenerator.genSuccessResultData(
            classService.list(pageQuery));
}

//查询所有班级信息列表
@RequestMapping("all")
@ResponseBody
public Response getAll(@RequestParam(required = false) Integer collegeId) {
    return ResultGenerator.genSuccessResultData(
            classService.getAll(collegeId));
}

//查询班级信息
@RequestMapping(path = "/{id}", method = {RequestMethod.GET})
@ResponseBody
public Response getById(@PathVariable Integer id) {
    return ResultGenerator.genSuccessResultData(
            classService.getById(id));
}

//添加班级信息
@RequestMapping(path = "/add", method = {RequestMethod.POST})
@ResponseBody
public Response addClass(@RequestBody Class theClass) {
    return ResultGenerator.genSuccessResultData(
            classService.add(theClass));
}

//修改班级信息
@RequestMapping(path = "/update", method = {RequestMethod.PUT})
@ResponseBody
public Response updateClass(@RequestBody Class theClass) {
    return ResultGenerator.genSuccessResultData(
            classService.update(theClass));
}

//删除班级信息
@RequestMapping(path = "/{id}/delete", method = {RequestMethod.DELETE})
@ResponseBody
public Response deleteClass(@PathVariable Integer id) {
    return ResultGenerator.genSuccessResultData(
            classService.delete(id));
}
}
```

### 3. 测试功能

在 10.6.1 节中已经详细测试了学院管理的各个页面和功能，班级管理与学院管理的删除功能基本一致，并且添加班级与修改班级页面也基本一样，因此，这里只介绍班级列表和添加班级功能。

(1) 启动前端系统和后端系统并访问前端系统，登录系统。

(2) 在页面左侧访问"班级管理"→"班级列表"，即可查看班级列表，如图 10-29 所示。

图 10-29　查看班级列表页面

(3) 单击系统左侧班级管理列表中的"添加班级"命令或者单击班级列表数据上面的"添加班级"按钮，进入添加班级页面，输入班级名称并选择对应的学院单击"立即添加"按钮，即可添加班级数据，如图 10-30 所示。

图 10-30　添加班级页面

## 10.6.3　学生管理

学生管理功能包含学生信息的列表查询、添加、修改和删除。

### 1. 开发前端

(1) 创建学生列表页面。

在 school-admin-ui 的 src/views 文件夹下创建 Student.vue 文件，代码如下：

```
<template>
```

```html
    <el-card class="good-container">
      <template #header>
        <div class="header">
          <el-button type="primary" :icon="Plus" @click="handleAdd">添加学生</el-button>
        </div>
      </template>
      <el-table :data="state.tableData" tooltip-effect="dark" style="width: 100%">
        <el-table-column prop="id" label="编号">
        </el-table-column>
        <el-table-column prop="name" label="姓名">
        </el-table-column>
        <el-table-column prop="serialNo" label="学号">
        </el-table-column>
        <el-table-column prop="clazz.name" label="班级">
        </el-table-column>
        <el-table-column prop="clazz.college.name" label="学院">
        </el-table-column>

        <el-table-column label="操作" width="100">
          <template #default="scope">
            <a style="cursor: pointer; margin-right: 10px" @click="handleEdit(scope.row.id)">修改</a>
            <el-popconfirm
              title="确定删除吗？"
              confirmButtonText='确定'
              cancelButtonText='取消'
              @confirm="handleDelete(scope.row.id)"
            >
              <template #reference>
                <a style="cursor: pointer">删除</a>
              </template>
            </el-popconfirm>
          </template>
        </el-table-column>
      </el-table>
      <!--总数超过一页，则展示分页器-->
      <el-pagination v-model:current-page="state.currentPage" v-model:page-size="state.pageSize" :page-sizes="[10, 20, 50, 100]"
          :small="small" :disabled="disabled" :background="background" layout="sizes, prev, pager, next" :total="state.total"
          @size-change="changeSize" @current-change="changePage" />
    </el-card>
</template>

<script setup>
import axios from '@/utils/axios'
import { Plus } from '@element-plus/icons-vue'
import { ElMessage } from 'element-plus'
import { getCurrentInstance, onMounted, onUnmounted, reactive } from 'vue'
import { useRouter } from 'vue-router'
import { localGet, localRemove, localSet } from '../utils'
```

```js
const app = getCurrentInstance()
const { goTop } = app.appContext.config.globalProperties
const router = useRouter()
const state = reactive({
  loading: false,
  tableData: [], // 数据列表
  total: 0, // 总条数
  currentPage: 1, // 当前页
  pageSize: 10 // 分页大小
})

onMounted(() => {
  state.currentPage=localGet("currentPage")===null?1:localGet("currentPage")
  state.pageSize=localGet("pageSize")===null?10:localGet("pageSize")
  getStudentList()
})
onUnmounted(() => {
  localRemove("currentPage");
  localRemove("pageSize");
});
const getStudentList = () => {
  state.loading = true
  axios.get('/student/list', {
    params: {
      pageNumber: state.currentPage,
      pageSize: state.pageSize
    }
  }).then(res => {
    state.tableData = res.data.data.list
    state.total = res.data.data.total
    state.currentPage = res.data.data.currentPage
    state.loading = false
    goTop && goTop()
  })
}
const handleAdd = () => {
  router.push({ path: '/addStudent' })
}
const handleEdit = (id) => {
  router.push({ path: '/addStudent', query: { id } })
}
const handleDelete = (id) => {
  axios.delete(`/student/${id}/delete`, {
  }).then(() => {
    ElMessage.success('删除成功')
    getStudentList()
  })
}
const changePage = (val) => {
  state.currentPage = val
  localSet("currentPage",val)
  getStudentList()
```

```
}
const changeSize = (val) => {
  state.pageSize = val
  localSet("pageSize",val)
  getStudentList()
}

</script>

<style scoped>
.good-container {
  min-height: 100%;
}

.el-card.is-always-shadow {
  min-height: 100% !important;
}
</style>
```

(2) 创建添加学生页面。

在 school-admin-ui 的 src/views 文件夹下创建 AddStudent.vue 文件，代码如下：

```
<template>
  <div class="add">
    <el-card class="add-container">
      <el-form :model="state.studentForm" :rules="state.rules" ref="studentRef" label-width="100px" class="classForm">
        <el-form-item label="姓名" prop="name">
          <el-input style="width: 300px" v-model="state.studentForm.name" placeholder="姓名"></el-input>
        </el-form-item>
        <el-form-item label="班级" prop="class">
          <el-select v-model="state.classValue" value-key="name" class="m-2" style="width: 300px;" placeholder="班级" size="large">
            <el-option v-for="item in state.classList" :key="item.id" :label="item.name" :value="item" />
          </el-select>
        </el-form-item>
        <el-form-item label="性别" prop="gender">
          <el-select v-model="state.genderValue" value-key="name" class="m-2" style="width: 300px;" placeholder="性别" size="large">
            <el-option v-for="item in state.genderList" :key="item.name" :label="item.name" :value="item" />
          </el-select>
        </el-form-item>
        <el-form-item label="学号" prop="serialNo">
          <el-input style="width: 300px" v-model="state.studentForm.serialNo" placeholder="学号"></el-input>
        </el-form-item>
        <el-form-item label="年龄" prop="age">
          <el-input style="width: 300px" v-model="state.studentForm.age" placeholder="年龄"></el-input>
        </el-form-item>
```

```
          <el-form-item>
            <el-button type="primary" @click="submitAdd()">{{ state.id ? '立
即修改' : '立即添加' }}</el-button>
          </el-form-item>
        </el-form>
      </el-card>
    </div>
</template>

<script setup>
import axios from '@/utils/axios';
import { ElMessage } from 'element-plus';
import { onMounted, reactive, ref } from 'vue';
import { useRoute } from 'vue-router';

const studentRef = ref(null)
const route = useRoute()
const { id } = route.query
const state = reactive({
  id: id,
  studentForm: {
    name: '',
    classId: 0,
    serialNo: '',
    age: 0,
  },
  classList: [],
  genderList: [{ name: "male" }, { name: "female" }],
  classValue: {},
  genderValue: {},
  rules: {
    name: [
      { required: 'true', message: '请填写班级名称', trigger: ['change'] }
    ]
  },
})
onMounted(() => {

  axios.get(`/class/all`).then(res => {
    state.classList = res.data.data
  })
  if (id) {
    axios.get(`/student/${id}`).then(res => {
      state.studentForm = {
        name: res.data.data.name,
        classId: res.data.data.classId,
        className: res.data.data.className,
        gender: res.data.data.gender,
        age: res.data.data.age,
        serialNo: res.data.data.serialNo
      }

      state.genderValue = { name: res.data.data.gender }
```

```
        state.classValue = { id: res.data.data.clazz.id, name: res.data.
data.clazz.name }

    })
  }

})
const submitAdd = () => {
  studentRef.value.validate((vaild) => {
    if (vaild) {
      let params = {
        name: state.studentForm.name,
        serialNo: state.studentForm.serialNo,
        clazz:{
          id:state.classValue.id,
        },
        gender: state.genderValue.name,
        age: state.studentForm.age,
      }
      console.log('params', params)
      if (id) {
        params.id = id
        axios.put(`/student/update`, params).then(() => {
          ElMessage.success('修改成功')
        })
      } else {
        axios.post(`/student/add`, params).then(() => {
          ElMessage.success('添加成功')
        })
      }
    }
  })
}
</script>

<style scoped>
.add {
  display: flex;
}

.add-container {
  flex: 1;
  height: 100%;
}

.avatar-uploader {
  width: 100px;
  height: 100px;
  color: #ddd;
  font-size: 30px;
}

.avatar-uploader-icon {
```

```
    display: block;
    width: 100%;
    height: 100%;
    border: 1px solid #e9e9e9;
    padding: 32px 17px;
}
</style>
```

(3) 注册学生列表与添加页面。

在 school-admin-ui 的 src/router 文件夹下找到 index.js 文件,在文件的

```
const routes = []
```

中添加如下学生列表和添加学生两个页面:

```
{
  path: '/student',
  name: 'student',
  component: () => import('../views/Student.vue')
},
{
  path: '/addStudent',
  name: 'addStudent',
  component: () => import('../views/AddStudent.vue')
},
```

### 2. 开发后端

(1) 创建 student 表的持久化类。

在 school-admin 的 src/main/java 目录下的 com.demo.school.bean 包中创建数据表 student 的持久化类 Student,代码如下:

```
package com.demo.school.bean;

import java.io.Serializable;

public class Student implements Serializable {
    //学生id
    private Integer id;
    //学生姓名
    private String name;
    //学生学号
    private String serialNo;
    //学生性别
    private String gender;
    //学生年龄
    private Integer age;
    //学生班级
    private Class clazz;
    //getter 和 setter 方法省略
}
```

(2) 开发数据访问层。

在 school-admin 的 src/main/java 目录下的 com.demo.school.dao 包中创建数据访问接口 StudentDao，代码如下：

```java
package com.demo.school.dao;

import com.demo.school.bean.Student;
import com.demo.school.utils.PageQuery;
import org.apache.ibatis.annotations.Mapper;
import org.springframework.stereotype.Repository;

import java.util.List;

/*
学生信息数据访问接口
 */
@Repository
@Mapper
public interface StudentDao {
    //分页查询学生列表
    List<Student> list(PageQuery pageQuery);
    //查询所有学生列表
    List<Student> getAll(Integer collegeId);
    //查询所有学生总数
    Integer getTotal();
    //查询学生信息
    Student getById(Integer id);
    //添加学生信息
    Integer add(Student student);
    //修改学生信息
    Integer update(Student student);
    //删除学生信息
    Integer delete(Integer id);
}
```

在 school-admin 的 src/main/java/resources/mapper 文件夹下创建 StudentDao.xml 文件，并添加和 StudentDao 接口对应的 MyBatis 的数据库操作内容，学生 student 和班级 class 是一对一的关联关系，所以需要定义 resultMap 实现关联，代码如下：

```xml
<!DOCTYPE mapper
        PUBLIC "-//mybatis.org//DTD Mapper 3.0//EN"
        "https://mybatis.org/dtd/mybatis-3-mapper.dtd">
<mapper namespace="com.demo.school.dao.StudentDao">
    <resultMap id="studentResultMap" type="Student">
        <id property="id" column="id"/>
        <result property="name" column="name"/>
        <result property="gender" column="gender"/>
        <result property="age" column="age"/>
        <result property="serialNo" column="serial_no"/>
        <association property="clazz" javaType="Class">
            <id property="id" column="class_id"/>
```

```xml
            <result property="name" column="class_name"/>
            <association property="college" javaType="College">
                <id property="id" column="college_id"/>
                <result property="name" column="college_name"/>
            </association>
        </association>
    </resultMap>
    <select id="getById" parameterType="int" resultMap="studentResultMap">
        SELECT a.*,
        b.id as class_id,b.name as class_name,
        c.id as college_id,c.name as college_name
        FROM student a
        inner join class b on a.class_id=b.id
        inner join college c on b.college_id=c.id
        WHERE a.id=#{id};
    </select>
    <select id="list" parameterType="PageQuery" resultMap="studentResultMap">
        SELECT a.*,
        b.id as class_id,b.name as class_name,
        c.id as college_id,c.name as college_name
        FROM student a
        inner join class b on a.class_id=b.id
        inner join college c on b.college_id=c.id
        ORDER BY a.id asc limit #{start},#{pageSize};
    </select>
    <select id="getAll" resultType="Student">
        SELECT * FROM student WHERE 1=1
        <if test="classId!=null">AND class_id=#{classId}</if>
        ORDER BY id ASC;
    </select>
    <select id="getTotal" resultType="int">
        SELECT COUNT(*) FROM student;
    </select>
    <insert id="add" parameterType="Student">
        INSERT INTO
        student(name,serial_no,gender,age,class_id)
        VALUES(#{name},#{serialNo},#{gender},#{age},#{clazz.id});
    </insert>
    <update id="update" parameterType="Student">
        UPDATE student
        SET name=#{name},serial_no=#{serialNo},
        gender=#{gender},age=#{age},class_id=#{clazz.id}
        WHERE ID=${id};
    </update>
    <delete id="delete" parameterType="int">
        DELETE FROM student WHERE ID = #{id};
    </delete>
</mapper>
```

(3) 开发业务逻辑层。

在 school-admin 的 src/main/java 目录下的 com.demo.school.service 包中创建数据访问接

□ StudentService,代码如下:

```java
package com.demo.school.service;

import com.demo.school.bean.Student;
import com.demo.school.utils.PageQuery;
import com.demo.school.utils.PageResponse;

import java.util.List;

/*
学生信息 Service 接口
 */
public interface StudentService {
    //分页查询学生列表
    PageResponse list(PageQuery pageQuery);

    //查询学生信息
    Student getById(int id);

    //添加学生信息
    Integer add(Student student);

    //修改学生信息
    Integer update(Student student);

    //删除学生信息
    Integer delete(Integer id);

    //查询所有学生信息列表
    List<Student> getAll(Integer classId);
}
```

在 school-admin 的 src/main/java 目录下的 com.demo.school.service.impl 包中创建数据访问接口 StudentService 的实现类 StudentServiceImpl,代码如下:

```java
package com.demo.school.service.impl;

import com.demo.school.bean.Student;
import com.demo.school.dao.StudentDao;
import com.demo.school.service.StudentService;
import com.demo.school.utils.PageQuery;
import com.demo.school.utils.PageResponse;
import org.springframework.beans.factory.annotation.Autowired;
import org.springframework.stereotype.Service;

import java.util.List;

/*
学生信息 Service 实现类
 */
@Service
```

```java
public class StudentServiceImpl implements StudentService {

    //注入 StudentDao 对象
    @Autowired
    private StudentDao studentDao;

    //分页查询学生列表
    @Override
    public PageResponse list(PageQuery pageQuery) {
        List<Student> list = this.studentDao.list(pageQuery);
        int total = this.studentDao.getTotal();
        PageResponse pageResult = new PageResponse(
                list, total, pageQuery.getPageSize(), pageQuery.getPageNumber());
        return pageResult;
    }

    //查询学生信息
    @Override
    public Student getById(int id) {
        return studentDao.getById(id);
    }

    //添加学生信息
    @Override
    public Integer add(Student student) {
        return studentDao.add(student);
    }

    //修改学生信息
    @Override
    public Integer update(Student student) {
        return studentDao.update(student);
    }

    //删除学生信息
    @Override
    public Integer delete(Integer id) {
        return studentDao.delete(id);
    }

    //查询所有学生信息列表
    @Override
    public List<Student> getAll(Integer classId) {
        return this.studentDao.getAll(classId);
    }
}
```

(4) 开发请求处理层。

在 school-admin 的 src/main/java 目录下的 com.demo.school.controller 包中创建请求处理类 StudentController，其中的分页查询和响应结果使用 10.4 节中已经封装好的代码，代码

如下：

```java
package com.demo.school.controller;

import com.demo.school.bean.Student;
import com.demo.school.service.StudentService;
import com.demo.school.utils.PageQuery;
import com.demo.school.utils.Response;
import com.demo.school.utils.ResultGenerator;
import org.springframework.beans.factory.annotation.Autowired;
import org.springframework.stereotype.Controller;
import org.springframework.web.bind.annotation.*;

import java.util.HashMap;
import java.util.Map;

/*
学生信息Controller
 */
@Controller
@RequestMapping("/student")
@CrossOrigin
public class StudentController {
    //注入StudentService对象
    @Autowired
    private StudentService studentService;

    //分页查询学生列表
    @RequestMapping("list")
    @ResponseBody
    public Response getStudentList(@RequestParam(required = false) Integer pageNumber,
                                   @RequestParam(required = false) Integer pageSize) {
        //验证分页参数，页码pageNumber和每页数据数pageSize必须提供
        //pageNumber最小为1，pageSize最小为10
        if (pageNumber == null || pageNumber < 1
                || pageSize == null || pageSize < 10) {
            //请求中如果分页参数有错误则返回BadRequest
            return ResultGenerator.genFailResult(ResultGenerator.RESULT_CODE_BAD_REQUEST_ERROR, "分页参数异常！");
        }
        //创建分页查询参数
        PageQuery pageQuery = new PageQuery(pageNumber, pageSize);
        //调用分页查询并发挥成功结果
        return ResultGenerator.genSuccessResultData(
                studentService.list(pageQuery));
    }

    //查询所有学生信息列表
    @RequestMapping("all")
    @ResponseBody
    public Response getAll(@RequestParam(required = false) Integer classId)
```

```
{
    return ResultGenerator.genSuccessResultData(
            studentService.getAll(classId));
}

//查询学生信息
@RequestMapping(path = "/{id}", method = {RequestMethod.GET})
@ResponseBody
public Response getById(@PathVariable Integer id) {
    return ResultGenerator.genSuccessResultData(
            studentService.getById(id));
}

//添加学生信息
@RequestMapping(path = "/add", method = {RequestMethod.POST})
@ResponseBody
public Response addStudent(@RequestBody Student student) {
    return ResultGenerator.genSuccessResultData(
            studentService.add(student));
}

//修改学生信息
@RequestMapping(path = "/update", method = {RequestMethod.PUT})
@ResponseBody
public Response updateStudent(@RequestBody Student student) {
    return ResultGenerator.genSuccessResultData(
            studentService.update(student));
}

//删除学生信息
@RequestMapping(path = "/{id}/delete", method = {RequestMethod.DELETE})
@ResponseBody
public Response deleteStudent(@PathVariable Integer id) {
    return ResultGenerator.genSuccessResultData(
            studentService.delete(id));
}
}
```

### 3. 测试功能

在 10.6.1 节中已经详细测试了学院管理的各个页面和功能，学生管理与学院管理的删除功能基本一致，并且添加学生与修改学生页面也基本一样，因此，这里只介绍学生列表和添加学生功能。

(1) 启动前端系统和后端系统并访问前端系统，登录系统。

(2) 在页面左侧访问"学生管理"→"学生列表"即可查看学生列表，如图 10-31 所示。

(3) 单击系统左侧菜单的学生管理列表中的"添加学生"命令或者单击学生列表数据上面的"添加学生"按钮，进入添加学生页面，输入学生姓名并选择对应的班级和性别，填写学号和年龄后单击"立即添加"按钮，即可添加学生数据，如图 10-32 所示。

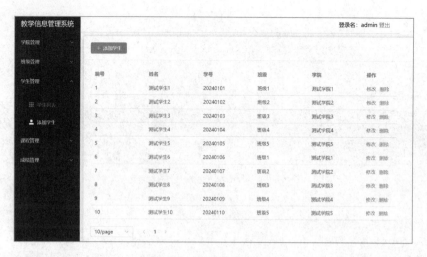

图 10-31　查看学生列表页面

图 10-32　添加学生页面

## 10.6.4　课程管理

课程管理功能包含课程信息的列表查询、添加、修改和删除。

### 1. 开发前端

(1) 创建课程列表页面。

在 school-admin-ui 的 src/views 文件夹下创建 Course.vue 文件，代码如下：

```
<template>
  <el-card class="good-container">
    <template #header>
      <div class="header">
        <el-button type="primary" :icon="Plus" @click="handleAdd">添加课程</el-button>
      </div>
    </template>
    <el-table :data="state.tableData" tooltip-effect="dark" style="width: 100%">
      <el-table-column prop="id" label="编号">
      </el-table-column>
      <el-table-column prop="name" label="课程名称">
```

```html
            </el-table-column>

            <el-table-column label="操作" width="100">
              <template #default="scope">
                <a style="cursor: pointer; margin-right: 10px" @click="handleEdit(scope.row.id)">修改</a>
                <el-popconfirm
                  title="确定删除吗？"
                  confirmButtonText='确定'
                  cancelButtonText='取消'
                  @confirm="handleDelete(scope.row.id)"
                >
                  <template #reference>
                    <a style="cursor: pointer">删除</a>
                  </template>
                </el-popconfirm>
              </template>
            </el-table-column>
        </el-table>
        <!--总数超过一页，则展示分页器-->
        <el-pagination v-model:current-page="state.currentPage" v-model:page-size="state.pageSize" :page-sizes="[10, 20, 50, 100]"
             :small="small"   :disabled="disabled"   :background="background"
layout="sizes, prev, pager, next" :total="state.total"
             @size-change="changeSize" @current-change="changePage" />
    </el-card>
</template>
```

```javascript
<script setup>
import axios from '@/utils/axios'
import { Plus } from '@element-plus/icons-vue'
import { ElMessage } from 'element-plus'
import { getCurrentInstance, onMounted, onUnmounted, reactive } from 'vue'
import { useRouter } from 'vue-router'
import { localGet, localRemove, localSet } from '../utils'

const app = getCurrentInstance()
const { goTop } = app.appContext.config.globalProperties
const router = useRouter()
const state = reactive({
  loading: false,
  tableData. [], // 数据列表
  total: 0, // 总条数
  currentPage: 1, // 当前页
  pageSize: 10 // 分页大小
})

onMounted(() => {
    state.currentPage=localGet("currentPage")===null?1:localGet("currentPage")
    state.pageSize=localGet("pageSize")===null?10:localGet("pageSize")
    getCourseList()
})
```

```javascript
onUnmounted(() => {
  localRemove("currentPage");
  localRemove("pageSize");
});
const getCourseList = () => {
  state.loading = true
  axios.get('/course/list', {
    params: {
      pageNumber: state.currentPage,
      pageSize: state.pageSize
    }
  }).then(res => {
    state.tableData = res.data.data.list
    state.total = res.data.data.total
    state.currentPage = res.data.data.currentPage
    state.loading = false
    goTop && goTop()
  })
}
const handleAdd = () => {
  router.push({ path: '/addCourse' })
}
const handleEdit = (id) => {
  router.push({ path: '/addCourse', query: { id } })
}
const handleDelete = (id) => {
  axios.delete(`/course/${id}/delete`, {
  }).then(() => {
    ElMessage.success('删除成功')
    getCourseList()
  })
}
const changePage = (val) => {
  state.currentPage = val
  localSet("currentPage",val)
  getCourseList()
}
const changeSize = (val) => {
  state.pageSize = val
  localSet("pageSize",val)
  getCourseList()
}
</script>

<style scoped>
.good-container {
  min-height: 100%;
}

.el-card.is-always-shadow {
  min-height: 100% !important;
}
</style>
```

(2) 创建添加课程页面。

在 school-admin-ui 的 src/views 文件夹下创建 AddCourse.vue 文件，代码如下：

```vue
<template>
  <div class="add">
    <el-card class="add-container">
      <el-form :model="state.courseForm" :rules="state.rules" ref="courseRef" label-width="100px" class="courseForm">
        <el-form-item label="课程" prop="course">
          <el-input style="width: 300px" v-model="state.courseForm.name" placeholder="请输课程名称"></el-input>
        </el-form-item>
        <el-form-item>
          <el-button type="primary" @click="submitAdd()">{{ state.id ? '立即修改' : '立即添加' }}</el-button>
        </el-form-item>
      </el-form>
    </el-card>
  </div>
</template>

<script setup>
import { reactive, ref, onMounted } from 'vue'
import axios from '@/utils/axios'
import { ElMessage } from 'element-plus'
import { useRoute } from 'vue-router'

const courseRef = ref(null)
const route = useRoute()
const { id } = route.query
const state = reactive({
  id: id,
  courseForm: {
    name: ''
  },
  rules: {
    name: [
      { required: 'true', message: '请填写课程名称', trigger: ['change'] }
    ]
  },
})
onMounted(() => {
  if (id) {
    axios.get(`/course/${id}`).then(res => {
      state.courseForm = {
        name: res.data.data.name,

      }

    })
  }
})
const submitAdd = () => {
```

```
      courseRef.value.validate((vaild) => {
        if (vaild) {
          let params = {
            name: state.courseForm.name
          }
          console.log('params', params)
          if (id) {
            params.id = id
            axios.put(`/course/update`, params).then(() => {
              ElMessage.success('修改成功')
            })
          } else {
            axios.post(`/course/add`, params).then(() => {
              ElMessage.success('添加成功')
            })
          }
        }
      })
    }
</script>

<style scoped>
.add {
  display: flex;
}

.add-container {
  flex: 1;
  height: 100%;
}

.avatar-uploader {
  width: 100px;
  height: 100px;
  color: #ddd;
  font-size: 30px;
}

.avatar-uploader-icon {
  display: block;
  width: 100%;
  height: 100%;
  border: 1px solid #e9e9e9;
  padding: 32px 17px;
}
</style>
```

(3) 注册课程列表与添加页面。

在 school-admin-ui 的 src/router 文件夹下找到 index.js 文件，在文件的

```
const routes = []
```

中添加如下课程列表和添加课程两个页面：

```
  {
    path: '/course',
    name: 'course',
    component: () => import('../views/Course.vue')
  },
  {
    path: '/addCourse',
    name: 'addCourse',
    component: () => import('../views/AddCourse.vue')
  },
```

**2. 开发后端**

（1）创建 course 表的持久化类。

在 school-admin 的 src/main/java 目录下的 com.demo.school.bean 包中创建数据表 course 的持久化类 Course，代码如下：

```
package com.demo.school.bean;

import java.io.Serializable;

public class Course implements Serializable {
    //课程 id
    private Integer id;
    //课程名称
    private String name;
    //getter 和 setter 方法省略
}
```

（2）开发数据访问层。

在 school-admin 的 src/main/java 目录下的 com.demo.school.dao 包中创建数据访问接口 CourseDao，代码如下：

```
package com.demo.school.dao;

import com.demo.school.bean.Course;
import com.demo.school.utils.PageQuery;
import org.apache.ibatis.annotations.Mapper;
import org.springframework.stereotype.Repository;

import java.util.List;

/*
课程信息数据访问接口
 */
@Repository
@Mapper
public interface CourseDao {
    //分页查询课程列表
    List<Course> list(PageQuery pageQuery);
```

```java
    //查询所有课程列表
    List<Course> getAll();

    //查询所有课程总数
    Integer getTotal();

    //查询课程信息
    Course getById(Integer id);

    //添加课程信息
    Integer add(Course course);

    //修改课程信息
    Integer update(Course course);

    //删除课程信息
    Integer delete(Integer id);
}
```

在 school-admin 的 src/main/java/resources/mapper 文件夹下创建 CourseDao.xml 文件，并添加和 CourseDao 接口对应的 MyBatis 的数据库操作内容，代码如下：

```xml
<!DOCTYPE mapper
        PUBLIC "-//mybatis.org//DTD Mapper 3.0//EN"
        "https://mybatis.org/dtd/mybatis-3-mapper.dtd">
<mapper namespace="com.demo.school.dao.CourseDao">
    <select id="getById" parameterType="int" resultType="Course">
        SELECT * FROM course WHERE id=#{id};
    </select>
    <select id="list" parameterType="PageQuery" resultType="Course">
        SELECT * FROM course
        ORDER BY id ASC LIMIT #{start},#{pageSize};
    </select>
    <select id="getAll" resultType="Course">
        SELECT * FROM course ORDER BY id ASC;
    </select>
    <select id="getTotal" resultType="int">
        SELECT COUNT(*) FROM course;
    </select>
    <insert id="add" parameterType="Course">
        INSERT INTO course(name) VALUES(#{name});
    </insert>
    <update id="update" parameterType="Course">
        UPDATE course SET name=#{name} WHERE id=${id};
    </update>
    <delete id="delete" parameterType="int">
        DELETE FROM course WHERE id = #{id};
    </delete>
</mapper>
```

(3) 开发业务逻辑层。

在 school-admin 的 src/main/java 目录下的 com.demo.school.service 包中创建数据访问接

口 CourseService,代码如下:

```java
package com.demo.school.service;

import com.demo.school.bean.Course;
import com.demo.school.utils.PageQuery;
import com.demo.school.utils.PageResponse;

import java.util.List;

/*
课程信息 Service 接口
 */
public interface CourseService {
    //分页查询课程列表
    PageResponse list(PageQuery pageQuery);

    //查询课程信息
    Course getById(int id);

    //添加课程信息
    Integer add(Course course);

    //修改课程信息
    Integer update(Course course);

    //删除课程信息
    Integer delete(Integer id);

    //查询所有课程信息列表
    List<Course> getAll();
}
```

在 school-admin 的 src/main/java 目录下的 com.demo.school.service.impl 包中创建数据访问接口 CourseService 的实现类 CourseServiceImpl,代码如下:

```java
package com.demo.school.service.impl;

import com.demo.school.bean.Course;
import com.demo.school.dao.CourseDao;
import com.demo.school.service.CourseService;
import com.demo.school.utils.PageQuery;
import com.demo.school.utils.PageResponse;
import org.springframework.beans.factory.annotation.Autowired;
import org.springframework.stereotype.Service;

import java.util.List;

/*
课程信息 Service 实现类
 */
@Service
```

```java
public class CourseServiceImpl implements CourseService {

    //注入CourseDao对象
    @Autowired
    private CourseDao courseDao;

    //分页查询课程列表
    @Override
    public PageResponse list(PageQuery pageQuery) {
        List<Course> list = this.courseDao.list(pageQuery);
        int total = this.courseDao.getTotal();
        PageResponse pageResult = new PageResponse(
                list, total, pageQuery.getPageSize(), pageQuery.getPageNumber());
        return pageResult;
    }

    //查询课程信息
    @Override
    public Course getById(int id) {
        return courseDao.getById(id);
    }

    //添加课程信息
    @Override
    public Integer add(Course course) {
        return courseDao.add(course);
    }

    //修改课程信息
    @Override
    public Integer update(Course course) {
        return courseDao.update(course);
    }

    //删除课程信息
    @Override
    public Integer delete(Integer id) {
        return courseDao.delete(id);
    }

    //查询所有课程信息列表
    @Override
    public List<Course> getAll() {
        return this.courseDao.getAll();
    }
}
```

(4) 开发请求处理层。

在school-admin的src/main/java目录下的com.demo.school.controller包中创建请求处理类CourseController,其中的分页查询和响应结果使用10.4节中已经封装好的代码,代码如下:

```java
package com.demo.school.controller;

import com.demo.school.bean.Course;
import com.demo.school.service.CourseService;
import com.demo.school.utils.PageQuery;
import com.demo.school.utils.Response;
import com.demo.school.utils.ResultGenerator;
import org.springframework.beans.factory.annotation.Autowired;
import org.springframework.stereotype.Controller;
import org.springframework.web.bind.annotation.*;

import java.util.HashMap;
import java.util.Map;

/*
课程信息Controller
 */
@Controller
@RequestMapping("/course")
@CrossOrigin
public class CourseController {
    // 注入CourseService对象
    @Autowired
    private CourseService courseService;

    //分页查询课程列表
    @RequestMapping("list")
    @ResponseBody
    public Response getCourseList(@RequestParam(required = false) Integer pageNumber,
                                  @RequestParam(required = false) Integer pageSize) {
        //验证分页参数，页码pageNumber和每页数据数pageSize必须提供
        //pageNumber最小为1，pageSize最小为10
        if (pageNumber == null || pageNumber < 1
                || pageSize == null || pageSize < 10) {
            //请求中如果分页参数有错误则返回BadRequest
            return ResultGenerator.genFailResult(ResultGenerator.RESULT_CODE_BAD_REQUEST_ERROR, "分页参数异常！");
        }
        //创建分页查询参数
        PageQuery pageQuery = new PageQuery(pageNumber, pageSize);
        //调用分页查询并发挥成功结果
        return ResultGenerator.genSuccessResultData(
                courseService.list(pageQuery));
    }

    //查询所有课程信息列表
    @RequestMapping("all")
    @ResponseBody
    public Response getAll() {
        return ResultGenerator.genSuccessResultData(
                courseService.getAll());
```

```java
    }

    //查询课程信息
    @RequestMapping(path = "/{id}", method = {RequestMethod.GET})
    @ResponseBody
    public Response getById(@PathVariable Integer id) {
        return ResultGenerator.genSuccessResultData(
                courseService.getById(id));
    }

    //添加课程信息
    @RequestMapping(path = "/add", method = {RequestMethod.POST})
    @ResponseBody
    public Response addCourse(@RequestBody Course course) {
        return ResultGenerator.genSuccessResultData(
                courseService.add(course));
    }

    //修改课程信息
    @RequestMapping(path = "/update", method = {RequestMethod.PUT})
    @ResponseBody
    public Response updateCourse(@RequestBody Course course) {
        return ResultGenerator.genSuccessResultData(
                courseService.update(course));
    }

    //删除课程信息
    @RequestMapping(path = "/{id}/delete", method = {RequestMethod.DELETE})
    @ResponseBody
    public Response deleteCourse(@PathVariable Integer id) {
        return ResultGenerator.genSuccessResultData(
                courseService.delete(id));
    }
}
```

**3．测试功能**

在 10.6.1 节中已经详细测试了学院管理的各个页面和功能，课程管理与学院管理的删除功能基本一致，并且添加课程与修改课程页面也基本一样，因此，这里只介绍课程列表和添加课程功能。

（1）启动前端系统和后端系统并访问前端系统，登录系统。

（2）在页面左侧访问"课程管理"→"课程列表"即可查看课程列表，如图 10-33 所示。

（3）单击系统左侧课程管理列表中的"添加课程"命令或者单击课程列表数据上面的"添加课程"按钮，进入添加课程页面，输入课程名称并单击"立即添加"按钮，即可添加课程数据，如图 10-34 所示。

图 10-33  查看课程列表页面

图 10-34  添加课程页面

## 10.6.5  成绩管理

成绩管理功能包含成绩信息的列表查询、添加、修改和删除。

1. 开发前端

（1）创建成绩列表页面。

在 school-admin-ui 的 src/views 文件夹下创建 Score.vue 文件，代码如下：

```
<template>
  <el-card class="good-container">
    <template #header>
      <div class="header">
        <el-button type="primary" :icon="Plus" @click="handleAdd">添加成绩</el-button>
      </div>
    </template>
    <el-table :data="state.tableData" tooltip-effect="dark" style="width: 100%">
```

```html
            <el-table-column prop="id" label="编号">
            </el-table-column>
            <el-table-column prop="student.clazz.college.name" label="学院">
            </el-table-column>
            <el-table-column prop="student.clazz.name" label="班级">
            </el-table-column>
            <el-table-column prop="student.name" label="姓名">
            </el-table-column>
            <el-table-column prop="course.name" label="课程">
            </el-table-column>
            <el-table-column prop="score" label="分数">
            </el-table-column>

            <el-table-column label="操作" width="100">
              <template #default="scope">
                <a style="cursor: pointer; margin-right: 10px" @click="handleEdit(scope.row.id)">修改</a>
                <el-popconfirm
                    title="确定删除吗？"
                    confirmButtonText='确定'
                    cancelButtonText='取消'
                    @confirm="handleDelete(scope.row.id)"
                >
                  <template #reference>
                    <a style="cursor: pointer">删除</a>
                  </template>
                </el-popconfirm>
              </template>
            </el-table-column>
          </el-table>
          <!--总数超过一页，再展示分页器-->
          <el-pagination v-model:current-page="state.currentPage" v-model:page-size="state.pageSize" :page-sizes="[10, 20, 50, 100]"
              :small="small" :disabled="false" :background="background" :hide-on-single-page=false layout="sizes, prev, pager, next" :total="state.total"
              @size-change="changeSize" @current-change="changePage" />

  </el-card>
</template>

<script setup>
import axios from '@/utils/axios'
import { Plus } from '@element-plus/icons-vue'
import { ElMessage } from 'element-plus'
import { getCurrentInstance, onMounted, onUnmounted, reactive } from 'vue'
import { useRouter } from 'vue-router'
import { localGet, localRemove, localSet } from '../utils'

const app = getCurrentInstance()
const { goTop } = app.appContext.config.globalProperties
const router = useRouter()
const state = reactive({
  loading: false,
```

```js
      tableData: [], // 数据列表
      total: 0, // 总条数
      currentPage: 1, // 当前页
      pageSize: 10 // 分页大小
    })
    onMounted(() => {
      state.currentPage=localGet("currentPage")===null?1:localGet("currentPage")
      state.pageSize=localGet("pageSize")===null?10:localGet("pageSize")
      getScoreList()
    })
    onUnmounted(() => {
      localRemove("currentPage");
      localRemove("pageSize");
    });

    const getScoreList = () => {
      state.loading = true
      axios.get('/score/list', {
        params: {
          pageNumber: state.currentPage,
          pageSize: state.pageSize
        }
      }).then(res => {
        state.tableData = res.data.data.list
        state.total = res.data.data.total
        state.currentPage = res.data.data.currentPage
        state.loading = false
        goTop && goTop()
      })
    }
    const handleAdd = () => {
      router.push({ path: '/addScore' })
    }
    const handleEdit = (id) => {
      router.push({ path: '/addScore', query: { id } })
    }
    const handleDelete = (id) => {
      axios.delete(`/score/${id}/delete`, {
      }).then(() => {
        ElMessage.success('删除成功')
        getScoreList()
      })
    }
    const changePage = (val) => {
      state.currentPage = val
      localSet("currentPage",val)
      getScoreList()
    }
    const changeSize = (val) => {
      state.pageSize = val
      localSet("pageSize",val)
      getScoreList()
    }
```

```
</script>

<style scoped>
.good-container {
  min-height: 100%;
}

.el-card.is-always-shadow {
  min-height: 100% !important;
}
</style>
```

(2) 创建添加成绩页面。

在 school-admin-ui 的 src/views 文件夹下创建 AddScore.vue 文件，代码如下：

```
<template>
  <div class="add">
    <el-card class="add-container">
      <el-form :model="state.scoreForm" :rules="state.rules" ref="scoreRef" label-width="100px" class="scoreForm">
        <el-form-item label="学院" prop="college">
          <el-select v-model="state.collegeValue" value-key="name" class="m-2" style="width: 300px;" placeholder="学院" size="large" @change="showClass" :disabled="id?true:false">
            <el-option v-for="item in state.collegeList" :key="item.id" :label="item.name" :value="item" />
          </el-select>
        </el-form-item>
        <el-form-item label="班级" prop="class">
          <el-select v-model="state.classValue" value-key="name" class="m-2" style="width: 300px;" placeholder="班级" size="large" @change="showStudent" :disabled="id?true:false">
            <el-option v-for="item in state.classList" :key="item.id" :label="item.name" :value="item" />
          </el-select>
        </el-form-item>
        <el-form-item label="学生" prop="student">
          <el-select v-model="state.studentValue" value-key="name" class="m-2" style="width: 300px;" placeholder="学生" size="large" :disabled="id?true:false">
            <el-option v-for="item in state.studentList" :key="item.id" :label="item.name" :value="item" />
          </el-select>
        </el-form-item>
        <el-form-item label="课程" prop="course">
          <el-select v-model="state.courseValue" value-key="name" class="m-2" style="width: 300px;" placeholder="课程" size="large" :disabled="id?true:false">
            <el-option v-for="item in state.courseList" :key="item.id" :label="item.name" :value="item" />
          </el-select>
        </el-form-item>
        <el-form-item label="成绩" prop="score">
```

```html
              <el-input style="width: 300px" v-model="state.scoreForm.score" placeholder="成绩"></el-input>
        </el-form-item>
        <el-form-item>
          <el-button type="primary" @click="submitAdd()">{{ state.id ? '立即修改' : '立即添加' }}</el-button>
        </el-form-item>
      </el-form>
    </el-card>
  </div>
</template>

<script setup>
import axios from '@/utils/axios';
import { ElMessage } from 'element-plus';
import {  onMounted, reactive, ref } from 'vue';
import { useRoute } from 'vue-router';

const scoreRef = ref(null)
const route = useRoute()
const { id } = route.query
const state = reactive({
  id: id,
  scoreForm: {
    studentId: 0,
    courseId: 0,
    score:0
  },
  collegeList: [],
  collegeValue: {},
  classList:[],
  classListValue:{},
  studentList:[],
  studentValue:{},
  courseList:[],
  courseValue:{},
  rules: {
    name: [
      { required: 'true', message: '请填写班级名称', trigger: ['change'] }
    ]
  },
})
onMounted(() => {
  if (id) {
    axios.get(`/score/${id}`).then(res => {
      state.collegeList=[{id:0,name:res.data.data.student.clazz.college.name}]
      state.classList=[{id:0,name:res.data.data.student.clazz.name}]
      state.studentList=[{id:res.data.data.student.id,name:res.data.data.student.name}]
      state.courseList=[{id:res.data.data.course.id,name:res.data.data.course.name}]

      state.collegeValue=state.collegeList[0]
```

```js
        state.classValue=state.classList[0]
        state.studentValue=state.studentList[0]
        state.courseValue=state.courseList[0]

        state.scoreForm.score=res.data.data.score
      })
    }else{
      axios.get(`/college/all`).then(res => {
        state.collegeList = res.data.data
      })
      axios.get(`/course/all`).then(res => {
        state.courseList = res.data.data
      })
    }
})
const showClass = (val) => {
  axios.get('/class/all/?collegeId='+val.id).then(res => {
    state.classList=res.data.data
  })
}
const showStudent = (val) => {
  axios.get('/student/all?classId='+val.id).then(res => {
    state.studentList=res.data.data
  })
}
const submitAdd = () => {
  scoreRef.value.validate((vaild) => {
    if (vaild) {
      let params = {
        student:{id: state.studentValue.id},
        course:{id: state.courseValue.id},
        score:state.scoreForm.score
      }
      console.log('params', params)
      if (id) {
        params.id = id
        axios.put(`/score/update`, params).then(() => {
          ElMessage.success('修改成功')
        })
      } else {
        axios.post(`/score/add`, params).then(() => {
          ElMessage.success('添加成功')
        })
      }
    }
  })
}
</script>

<style scoped>
.add {
```

```css
    display: flex;
}
.add-container {
  flex: 1;
  height: 100%;
}
.avatar-uploader {
  width: 100px;
  height: 100px;
  color: #ddd;
  font-size: 30px;
}
.avatar-uploader-icon {
  display: block;
  width: 100%;
  height: 100%;
  border: 1px solid #e9e9e9;
  padding: 32px 17px;
}
</style>
```

(3) 注册成绩列表与添加页面。

在 school-admin-ui 的 src/router 文件夹下找到 index.js 文件，在文件的

```
const routes = []
```

中添加如下成绩列表和添加成绩两个页面：

```js
{
  path: '/score',
  name: 'score',
  component: () => import('../views/Score.vue')
},
{
  path: '/addScore',
  name: 'addScore',
  component: () => import('../views/AddScore.vue')
},
```

2. 开发后端

(1) 创建 score 表的持久化类。

在 school-admin 的 src/main/java 目录下的 com.demo.school.bean 包中创建数据表 score 的持久化类 Score，代码如下：

```java
package com.demo.school.bean;

import java.io.Serializable;

public class Score implements Serializable {
    //成绩 id
```

```java
    private Integer id;
    //成绩关联的学生
    private Student student;
    //成绩关联的课程
    private Course course;
    //成绩的值
    private Integer score;
}
```

(2) 开发数据访问层。

在 school-admin 的 src/main/java 目录下的 com.demo.school.dao 包中创建数据访问接口 ScoreDao，代码如下：

```java
package com.demo.school.dao;

import com.demo.school.bean.Score;
import com.demo.school.utils.PageQuery;
import org.apache.ibatis.annotations.Mapper;
import org.springframework.stereotype.Repository;

import java.util.List;

/*
成绩信息数据访问接口
 */
@Repository
@Mapper
public interface ScoreDao {
    //分页查询成绩列表
    List<Score> list(PageQuery pageQuery);

    //查询所有成绩列表
    List<Score> getAll();

    //查询所有成绩总数
    Integer getTotal();

    //查询成绩信息
    Score getById(Integer id);

    //添加成绩信息
    Integer add(Score score);

    //修改成绩信息
    Integer update(Score score);

    //删除成绩信息
    Integer delete(Integer id);
}
```

在 school-admin 的 src/main/java/resources/mapper 文件夹下创建 ScoreDao.xml 文件，并

添加和 ScoreDao 接口对应的 MyBatis 的数据库操作内容，成绩表 score 中的 student_id 和 course_id 需要与 Student 类和 Course 类关联，所以需要定义 resultMap 实现关联，代码如下：

```xml
<!DOCTYPE mapper
    PUBLIC "-//mybatis.org//DTD Mapper 3.0//EN"
    "https://mybatis.org/dtd/mybatis-3-mapper.dtd">
<mapper namespace="com.demo.school.dao.ScoreDao">
    <resultMap id="scoreResultMap" type="Score">
        <id property="id" column="id"/>
        <result property="score" column="score"/>
        <association property="student" javaType="Student">
            <id property="id" column="student_id"/>
            <result property="name" column="student_name"/>
            <result property="serialNo" column="student_serial"/>
            <association property="clazz" javaType="Class">
                <id property="id" column="class_id"/>
                <result property="name" column="class_name"/>
                <association property="college" javaType="College">
                    <id property="id" column="college_id"/>
                    <result property="name" column="college_name"/>
                </association>
            </association>
        </association>
        <association property="course" javaType="Course">
            <id property="id" column="course_id"/>
            <result property="name" column="course_name"/>
        </association>
    </resultMap>
    <select id="getById" parameterType="int" resultMap="scoreResultMap">
        SELECT a.*,
        b.id as student_id,b.name AS student_name,b.serial_no as student_serial,
        c.id as course_id,c.name AS course_name,
        d.id as class_id,d.name AS class_name,
        e.id as college_id,e.name AS college_name
        FROM score a
        INNER JOIN student b ON a.student_id=b.id
        INNER JOIN course c ON a.course_id=c.id
        INNER JOIN class d ON b.class_id=d.id
        INNER JOIN college e ON d.college_id=e.id
        WHERE a.id=#{id};
    </select>
    <select id="list" parameterType="PageQuery" resultMap="scoreResultMap">
        SELECT a.*,
        b.id as student_id,b.name AS student_name,b.serial_no as student_serial,
        c.id as course_id,c.name AS course_name,
        d.id as class_id,d.name AS class_name,
        e.id as college_id,e.name AS college_name
        FROM score a
        INNER JOIN student b ON a.student_id=b.id
```

```xml
                INNER JOIN course c ON a.course_id=c.id
                INNER JOIN class d ON b.class_id=d.id
                INNER JOIN college e ON d.college_id=e.id
                ORDER BY a.id asc limit #{start},#{pageSize}
        </select>
        <select id="getAll" resultType="Score">
            SELECT * FROM score ORDER BY id ASC;
        </select>
        <select id="getTotal" resultType="int">
            SELECT COUNT(*) FROM score;
        </select>
        <insert id="add" parameterType="Score">
            INSERT INTO score(student_id,course_id,score)
            VALUES(#{student.id},#{course.id},#{score});
        </insert>
        <update id="update" parameterType="Score">
            UPDATE score SET
            student_id=#{student.id},course_id=#{course.id},score=#{score}
            WHERE ID=${id};
        </update>
        <delete id="delete" parameterType="int">
            DELETE FROM score WHERE ID = #{id};
        </delete>
</mapper>
```

(3) 开发业务逻辑层。

在school-admin的src/main/java目录下的com.demo.school.service包中创建数据访问接口ScoreService，代码如下：

```java
package com.demo.school.service;

import com.demo.school.bean.Score;
import com.demo.school.utils.PageQuery;
import com.demo.school.utils.PageResponse;

import java.util.List;

/*
成绩信息Service接口
*/
public interface ScoreService {
    //分页查询成绩列表
    PageResponse list(PageQuery pageQuery);

    //查询成绩信息
    Score getById(int id);

    //添加成绩信息
    Integer add(Score score);

    //修改成绩信息
```

```java
    Integer update(Score score);

    //删除成绩信息
    Integer delete(Integer id);

    //查询所有成绩信息列表
    List<Score> getAll();
}
```

在 school-admin 的 src/main/java 目录下的 com.demo.school.service.impl 包中创建数据访问接口 ScoreService 的实现类 ScoreServiceImpl，代码如下：

```java
package com.demo.school.service.impl;

import com.demo.school.bean.Score;
import com.demo.school.dao.ScoreDao;
import com.demo.school.service.ScoreService;
import com.demo.school.utils.PageQuery;
import com.demo.school.utils.PageResponse;
import org.springframework.beans.factory.annotation.Autowired;
import org.springframework.stereotype.Service;

import java.util.List;

/*
成绩信息 Service 实现类
 */
@Service
public class ScoreServiceImpl implements ScoreService {

    //注入 ScoreDao 对象
    @Autowired
    private ScoreDao scoreDao;

    //分页查询成绩列表
    @Override
    public PageResponse list(PageQuery pageQuery) {
        List<Score> list = this.scoreDao.list(pageQuery);
        int total = this.scoreDao.getTotal();
        PageResponse pageResult = new PageResponse(
                list, total, pageQuery.getPageSize(), pageQuery.getPageNumber());
        return pageResult;
    }

    //查询成绩信息
    @Override
    public Score getById(int id) {
        return scoreDao.getById(id);
    }

    //添加成绩信息
    @Override
    public Integer add(Score score) {
```

```java
        return scoreDao.add(score);
    }

    //修改成绩信息
    @Override
    public Integer update(Score score) {
        return scoreDao.update(score);
    }

    //删除成绩信息
    @Override
    public Integer delete(Integer id) {
        return scoreDao.delete(id);
    }

    //查询所有成绩信息列表
    @Override
    public List<Score> getAll() {
        return this.scoreDao.getAll();
    }
}
```

(4) 开发请求处理层。

在 school-admin 的 src/main/java 目录下的 com.demo.school.controller 包中创建请求处理类 ScoreController，其中的分页查询和响应结果使用 10.4 节中已经封装好的代码，代码如下：

```java
package com.demo.school.controller;

import com.demo.school.bean.Score;
import com.demo.school.service.ScoreService;
import com.demo.school.utils.PageQuery;
import com.demo.school.utils.Response;
import com.demo.school.utils.ResultGenerator;
import org.springframework.beans.factory.annotation.Autowired;
import org.springframework.stereotype.Controller;
import org.springframework.web.bind.annotation.*;

import java.util.HashMap;
import java.util.Map;

/*
成绩信息 Controller
 */
@Controller
@RequestMapping("/score")
@CrossOrigin
public class ScoreController {
    // 注入 ScoreService 对象
    @Autowired
    private ScoreService scoreService;

    //分页查询成绩列表
    @RequestMapping("list")
```

```java
    @ResponseBody
    public Response getScoreList(@RequestParam(required = false) Integer pageNumber,
                                 @RequestParam(required = false) Integer pageSize) {
        //验证分页参数,页码 pageNumber 和每页数据数 pageSize 必须提供
        //pageNumber 最小为 1,pageSize 最小为 10
        if (pageNumber == null || pageNumber < 1
                || pageSize == null || pageSize < 10) {
            //请求中如果分页参数有错误则返回 BadRequest
            return ResultGenerator.genFailResult(ResultGenerator.RESULT_CODE_BAD_REQUEST_ERROR, "分页参数异常!");
        }
        //创建分页查询参数
        PageQuery pageQuery = new PageQuery(pageNumber, pageSize);
        //调用分页查询并发挥成功结果
        return ResultGenerator.genSuccessResultData(
                scoreService.list(pageQuery));
    }

    //查询所有成绩信息列表
    @RequestMapping("all")
    @ResponseBody
    public Response getAll() {
        return ResultGenerator.genSuccessResultData(
                scoreService.getAll());
    }

    //查询成绩信息
    @RequestMapping(path = "/{id}", method = {RequestMethod.GET})
    @ResponseBody
    public Response getById(@PathVariable Integer id) {

        return ResultGenerator.genSuccessResultData(
                scoreService.getById(id));

    }

    //添加成绩信息
    @RequestMapping(path = "/add", method = {RequestMethod.POST})
    @ResponseBody
    public Response addScore(@RequestBody Score score) {
        return ResultGenerator.genSuccessResultData(
                scoreService.add(score));
    }

    //修改成绩信息
    @RequestMapping(path = "/update", method = {RequestMethod.PUT})
    @ResponseBody
    public Response updateScore(@RequestBody Score score) {
        return ResultGenerator.genSuccessResultData(
                scoreService.update(score));
    }

    //删除成绩信息
```

```
@RequestMapping(path = "/{id}/delete", method = {RequestMethod.DELETE})
@ResponseBody
public Response deleteScore(@PathVariable Integer id) {
    return ResultGenerator.genSuccessResultData(
        scoreService.delete(id));
}
}
```

### 3. 测试功能

在 10.6.1 节中已经详细测试了学院管理的各个页面和功能，成绩管理与学院管理的删除功能基本一致，并且添加成绩与修改成绩页面也基本一样，因此，这里只介绍成绩列表和添加成绩功能。

(1) 启动前端系统和后端系统并访问前端系统，登录系统。

(2) 在页面左侧访问"成绩管理"→"成绩列表"，即可查看成绩列表，如图 10-35 所示。

图 10-35　查看成绩列表页面

(3) 单击系统左侧成绩管理列表中的"添加成绩"命令或者单击成绩列表数据上面的"添加成绩"按钮，进入添加成绩页面。首先在"学院"下拉列表中选择成绩所属学生的学院，选中学院后"班级"列表中就只能选择选中学院下的所有班级，班级选中后"学生"列表中就只能选择选中班级下的所有学生，然后选择学生并选择课程，最后输入成绩，单击"立即添加"按钮，即可添加成绩数据，如图 10-36 所示。

图 10-36　添加成绩页面

## 本 章 小 结

在本章中，我们深入研究了基于 Spring、SpringMVC、SpringBoot 和 MyBatis 框架项目的开发过程，包括创建前端与后端项目、开发后端 Web 服务、数据库集成、用户身份验证、分页查询、前端界面集成等关键项。以上几个框架的整合使得开发过程更加高效和便捷。

## 课 后 习 题

**实操题**

本章所开发的项目的使用者是系统管理员，管理员拥有所有数据的增删改查权限。请根据本章项目的内容添加学生、教师使用的系统，学生可以修改自己的信息、选课和查看自己的成绩，教师可以修改自己的信息、建课、录入所建课程的成绩。

### 平衡发言权利与内容规范

假设你正在使用 SpringBoot 开发一个社交媒体平台。平台规定，用户不能发布含有暴力或其他违法信息的内容。这时，作为开发者，你的责任不仅是实现一个高性能、用户友好的平台，更重要的是，要确保平台不成为传播不良信息的渠道。

在技术层面，你可以使用 SpringBoot 的文本分析库，创建一个敏感词过滤器，自动屏蔽或报告不当内容。但同时，也需要与法律顾问和伦理专家合作，以确保过滤算法不侵犯到用户的发言权利和个人隐私。

通过这个案例，我们可以看到，技术开发并不是孤立的，它与社会责任、法律和伦理紧密相连。作为一名合格的软件工程师，我们不仅要有扎实的技术基础，还需要具备社会责任感和道德规范意识。

# 参 考 文 献

[1] Spring 框架官方参考文档. https://docs.spring.io/spring-framework/docs/5.3.25/reference/html/.
[2] Springboot 框架官方参考文档. https://docs.spring.io/spring-boot/docs/2.7.9/reference/html/.
[3] Mybatis 官方参考文档. https://mybatis.org/mybatis-3/.
[4] Vue.js 官方文档. https://vuejs.org/guide/introduction.html.
[5] ElementPlus 官方文档.https://element-plus-docs.bklab.cn/zh-CN/.
[6] 陈恒，李正光. SSM+Spring Boot+Vue.js 3 全栈开发从入门到实战(微课视频版)[M]. 北京：清华大学出版社，2022.
[7] 黑马程序员. JavaEE 企业级应用开发教程(Spring+Spring MVC+MyBatis)[M]. 2 版. 北京：人民邮电出版社，2021.
[8] 陈学明. Spring+Spring MVC+MyBatis 整合开发实战[M]. 北京：机械工业出版社，2020.